献给三年以来关心支持北川重建的人们
献给为北川新县城重建作出贡献的人们

Dedicated to those who have been caring and supporting the reconstruction of Beichuan.
And, to those who have been devoted to the reconstruction of the new county seat.

《建筑新北川》编辑委员会

Editorial Committee of *Building a New Beichuan*

建筑新北川

Building A New Beichuan

中国城市规划设计研究院
China Academy of Urban Planning & Design
中国建筑设计研究院
China Architecture Design & Research Group
编著

中国建筑工业出版社
China Architecture & Building Press

目录 Contents

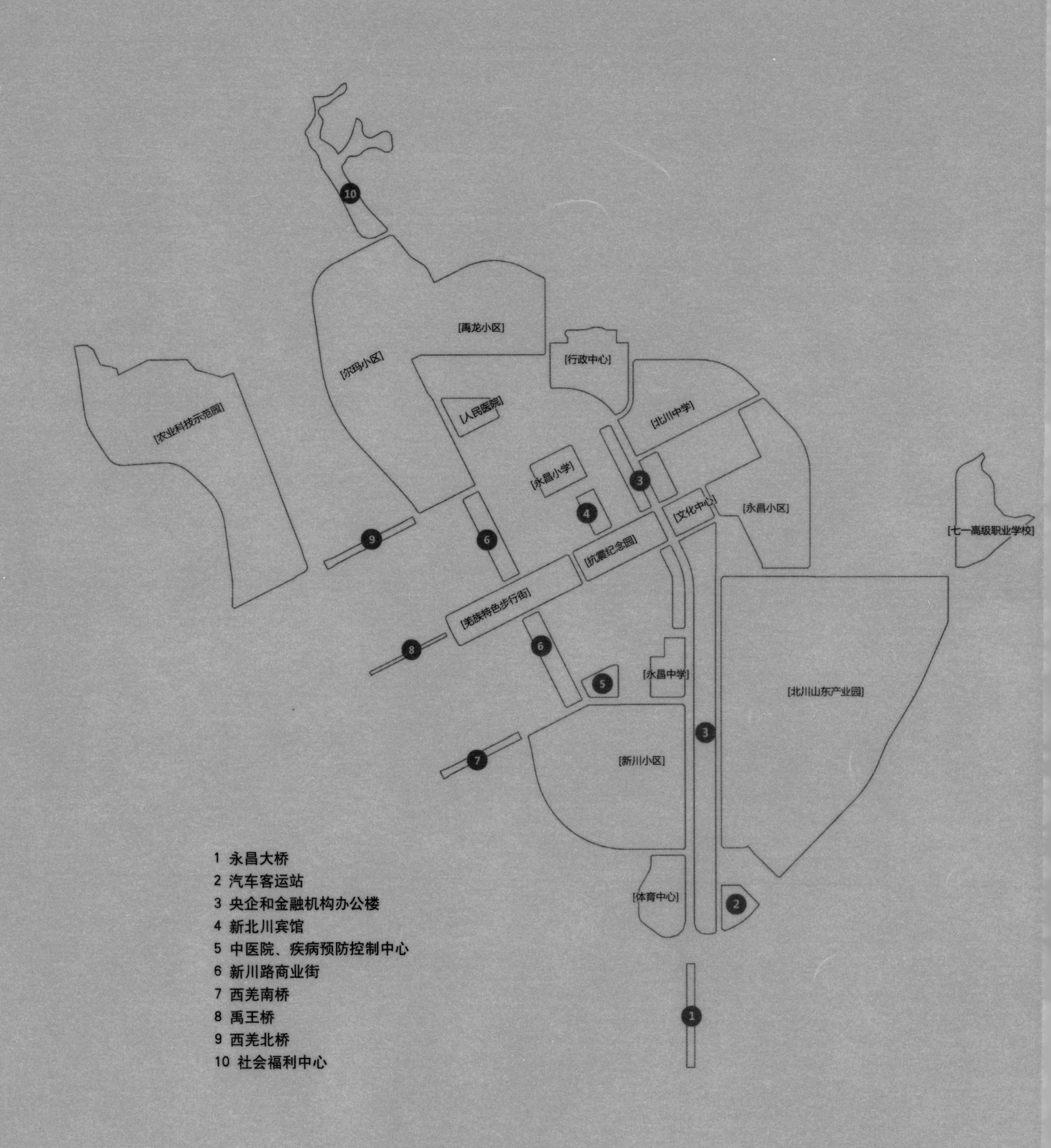
[禹龙小区]
[尔玛小区]
[行政中心]
[农业科技示范园]
[人民医院]
[北川中学]
[永昌小学]
[文化中心]
[永昌小区]
[七一高级职业学校]
[抗震纪念园]
[羌族特色步行街]
[永昌中学]
[北川山东产业园]
[新川小区]
[体育中心]
1 永昌大桥
2 汽车客运站
3 央企和金融机构办公楼
4 新北川宾馆
5 中医院、疾病预防控制中心
6 新川路商业街
7 西羌南桥
8 禹王桥
9 西羌北桥
10 社会福利中心

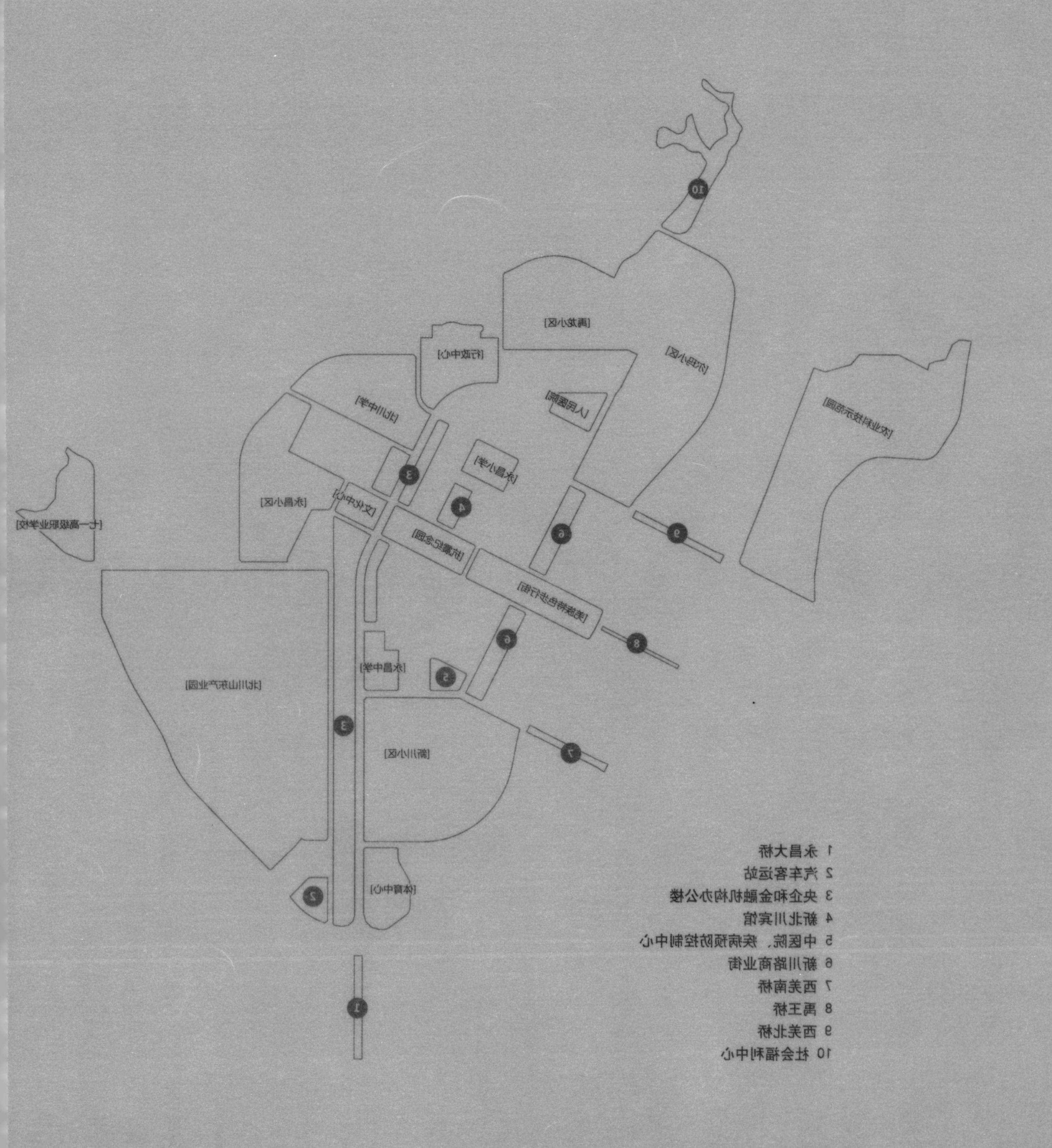

[七一高级职业学校]
[永昌小区]
[北川中学]
[行政中心]
[禹龙小区]
[尔玛小区]
[人民医院]
[永昌小学]
[农业科技示范园]
[文化中心]
[抗震纪念园]
[羌族特色步行街]
[北川山东产业园]
[永昌中学]
[新川小区]
[体育中心]
1 永昌大桥
2 汽车客运站
3 中央和金融机构办公楼
4 新北川宾馆
5 中医院、疾病预防控制中心
6 新川路商业街
7 西羌南桥
8 禹王桥
9 西羌北桥
10 社会福利中心

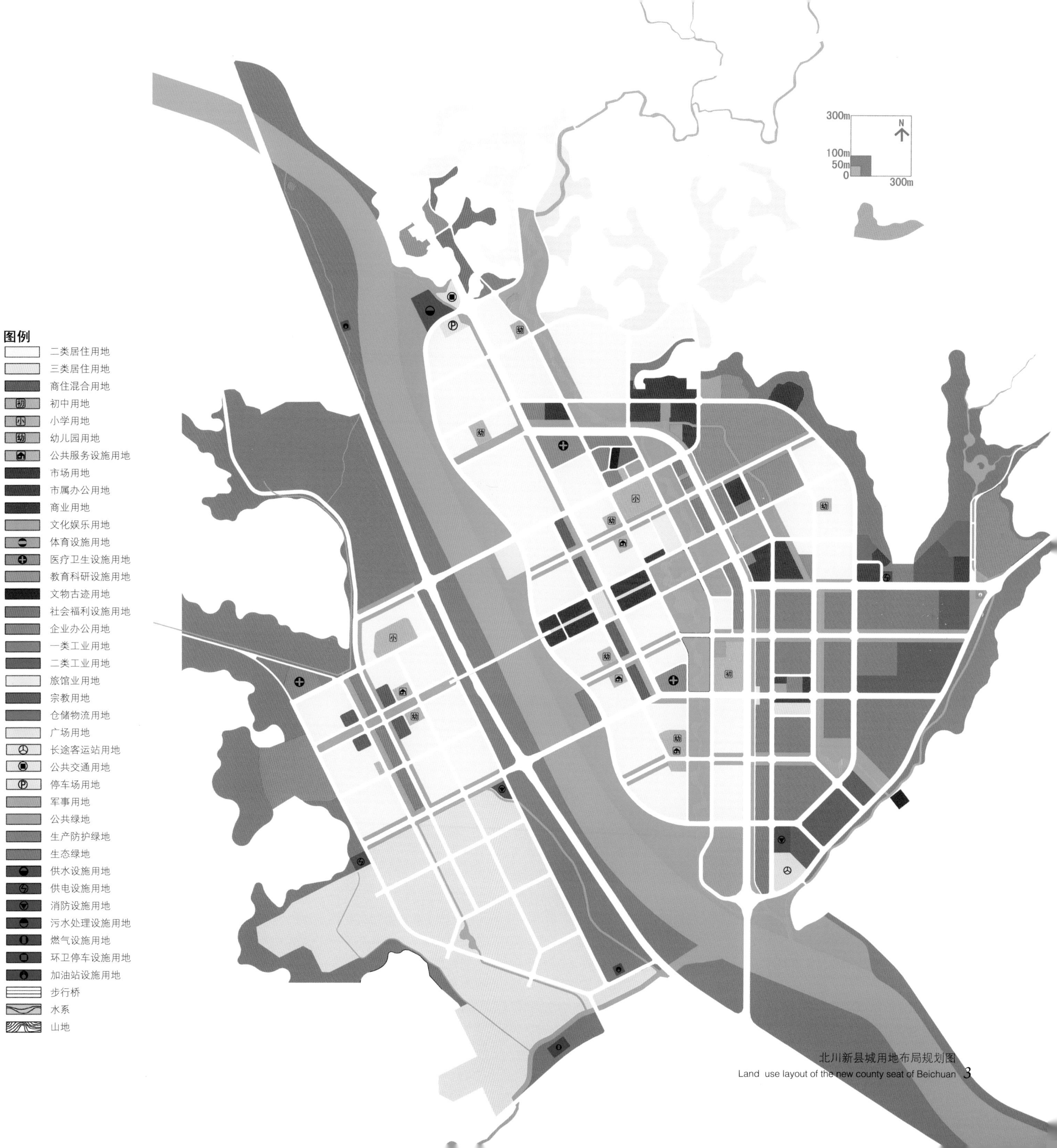

北川新县城用地布局规划图

Land use layout of the new county seat of Beichuan

前言·规划师与建筑师合作的典范

李晓江 / 中国城市规划设计研究院院长
中规院北川新县城规划工作指挥部指挥长

"5·12"汶川特大地震灾后重建，给了全国的规划师一个担当社会责任、以专业技术回馈社会的机会，也给了规划师与建筑师深度合作、相互学习和增进专业融通的难得机会！

2008年5月19日，当我带领住房和城乡建设部救灾应急规划绵阳组第一次走进北川老县城时，被灾难的惨烈深深地震撼；5月25日，第二次走进北川老县城，我们被北川县宋明书记、经大忠县长重建家园的迫切心情和他们对规划师的信任深深感动；5月底，中国城市规划设计研究院和武汉地质工程勘察院联合承担的《北川重建选址论证初步报告》仅用十天时间完成，6月初，报告经绵阳市委谭力书记签发，报送国务院抗震救灾总指挥部；6月3日，住建部仇保兴副部长考察北川老县城灾害情况和新县城选址，要求中规院尽全力做好北川灾后重建工作；8月，上任不久的北川县委陈兴春书记提出请中规院全面承担新县城重建的技术总协调，山东援川前线指挥部欣然接受了这一重要建议；10月，新县城规划工作正式启动；2009年10月，在绵阳市委吴靖平书记安排下，绵阳市北川新县城工程建设指挥部、中规院北川新县城规划工作指挥部先后成立，与山东援川前线指挥部共同形成了应对全面重建的"三个指挥部"联动的工作机制。陈兴春书记设想的规划设计技术"一个漏斗"的灾后重建全新模式由此诞生。

2008年11月16日，温家宝总理考察新县城选址时指出要把北川新县城建成"城建工程标志、抗震精神标志和文化遗产标志"，提出了新县城规划的"安全、宜居、特色、繁荣、和谐、文明"十二字方针。2009年5月11日，胡锦涛总书记视察新县城建设工地时指出"北川新县城建设是北川人民的企盼，是全国各族人民的期望"。虽然中规院几十年来承担过大量的重要规划，完成过综合复杂的实施项目，但面对北川新县城重建这样超乎寻常的标志性工程，我们实实在在体会到了胡总书记"责任重大、使命光荣"嘱托的含义，更感觉到了我们将要面对的工作难度和分量。

北川是我国唯一的羌族自治县，异地重建的县城和建筑如何传承、发展羌族历史文化和风貌特色；如何使建筑设计既能反映创意和水平，又使城市保持整体协调；如何能在很短的时间完成上百个、200多万平方米的建筑设计，是新县城技术总协调面临的最大挑战。

值得庆幸的是，北川的规划实施工作得到了诸多设计院和设计大师的热烈响应，尤其是得到中国建筑学会的鼎力支持。2009年6月28日，由中国建筑学会、绵阳市委市政府和中规院联合召开的"北川新县城城镇风貌和建筑风格专家研讨会"汇集了全国院士、设计大师、川西和羌族民居研究学者、地方文化和民俗专家等，会议达成的《关于北川新县城城镇风貌和建筑风格的六点共识》，成为北川新县城建筑风格与设计的重要指导文件，为北川新县城的建筑设计提供了关键的技术指导和评选标准。

建筑界德高望重的宋春华理事长和周畅秘书长不仅全力动员了一大批著名学者大师参与建筑方案设计与评审、遴选，更是多次亲临现场。为做到精益求精，不留遗憾，在北京、成都、绵阳和北川等地组织和主持了十多次会议。宋春华理事长带领的中国建筑学会在北川新县城建筑设计中作出的贡献是无可替代的！

张锦秋院士在新县城抗震纪念园方案难以选择的关键时候，亲自担任顾问，鼓励由规划师整合方案。在西安家中、办公室，在北京，在北川，大师以执着的精神、睿智的思想、温婉的态度，带领中规院设计小组不断深化规划构思，亲自动手修改方案。先生的悉心指导成就了北川新县城最重要的精神场所，成就了一个集多位大师智慧的集体作品。

崔愷大师承担文化中心的设计，第一轮方案就获得了一致好评。凭借着他对本土建筑传统及羌寨聚落的研究功底，用现代手法和当地材料创造了一个由表及里的全新"羌寨"，达成了外在形象和内部空间的高度统一。在无数次因为行政调整和面积、内容更改等原因的无奈修改中，大师以难能可贵的耐心和坚忍，尽力维护着设计理念。如今一座现代版的羌族山寨屹立在新北川文化景观中轴线的东端。

庄惟敏大师设计的幸福馆，是由原规划6000m^{2}的抗震纪念馆和城建档案馆压缩调整到2000m^{2}的建筑。庄大师在不断修改方案的过程中，既坚持初始的"白石"理念，又以谦和的态度配合整体规划的需要，匠心独具的建筑方案实现了与环境协调的个性和低调的张扬。

周恺大师设计的静思园是新县城极有感染力的场所，也许只有周大师作为唐山大地震幸存者特有的生命感悟，方能成就如此感人的作品。静思园是一个与逝者、与上天平和对话的场所，宁静中给人以冥想生命意义的深邃，平实中教人以思索自然力量的敬畏，实乃大师作品！

孟建民大师对北川新县城的倾情投入让我尤其感动。他不因事小而不为，坚持在纪念园英雄广场设计中"小题大做"，刻意设计了曲线伸展的地面铺装和地灯，营造了独特的视觉效果。真希望同行们有机会能在夜晚到北川新县城观赏孟大师的广场设计杰作。

香港华艺建筑设计有限公司的盛烨和陆强两位老总主动请求参与北川重建的设计工作。"请战"的态度和公司的实力，使他们在三大标志性建筑方案竞争中，成为唯一全部进入"三甲"的单位。陈日飚设计的行政中心方案，整体上大气庄重，不夸张、不浮华，追求羌族建筑文化的有机传承与创新，在尺度、比例、材料、色彩上精益求精。

清华大学的邓雪娴和西南交大的季富政两位老师带病完成的商业步行街和禹王桥的详细规划和建筑设计方案，具有浓郁的羌族传统风貌，成为新县城展示羌族建筑传统风貌和旅游的亮点。这种为民族文化复兴而忘我工作的精神，令所有援建者赞叹不已。

叶毓山大师作为著名的雕塑家，经宋春华理事长的力荐成为纪念园主题雕塑的作者。为了《新生》，叶大师不顾年事已高一次次到北京、北川参加会议，饱含激情地介绍主题构思，认真听取各方意见，创作了不同比例，直至1：1的模型来琢磨每一个细节。叶大师的《新生》既表达了中国人民战胜困难的英雄气概，又展示了羌族人民的美好生活前景，这座雅俗共赏的杰作成为新县城最核心的视觉焦点，宣示着一个民族、一座城市的新生。作为北川新县城的规划者，我特别感激叶大师倾情创作的《新生》为新县城"点睛"。

中国建筑设计研究院是北川新县城重建中最"给力"的建筑设计合作伙伴。修龙院长多次表示要做好中规院的技术后盾。刘燕辉书记带领的团队最早与中规院建筑设计团队合作，承担完成了新县城全部120万平方米住宅建筑的方案、初步设计与景观设计，保证新县城安置房成为最早开工、最先完工的建筑，实现了灾后重建民生优先的承诺。为了让所有的住宅建筑达到"绿色建筑"标准，刘燕辉一次次耐心说服建设单位，寻找降低成本的办法。重建后期，中建院又承担了大量"不起眼"，但与新县城功能完善密切相关的办公服务建筑设计。中建院设计的尔玛小区景观环境，融入了大量羌族文化元素，深受受灾群众的喜爱；永昌小学灵活的布局，院落式的空间组织和漂亮、文雅的外观，是对学校设计的创新探索，得到师生的好评。

中规院建筑设计所（北京国城建筑设计公司），虽然没有建筑大院的实力和大师的领衔，但在重建过程中虚心向大院学习、向大师学习，与中建院合作完成了全部住宅建筑的初步设计，并独立承担了大量住宅施工图、场地设计和一批公共建筑设计。他们是在北川新县城承担建筑设计工作量最大，现场配合施工、配合规划指挥部工作时间最长的单位。中规院建筑设计团队在北川的实践中也得到了很大的提升。

北川新县城的成果是政治家与规划师、建筑师相互理解、相互支持、精诚合作的范例，其中最典型的当属抗震纪念园。中国建筑学会帮助组织的三轮方案征集，产生了二十多个方案，其中不乏院士、大师的优秀创作。但三次评审都没有达成纪念园主题表达的统一意见。我与宋春华理事长商议，莫如在温家宝总理视察新县城现场时直接请示。2009年9月25日，在向温总理汇报新县城规划设计时，我提到了专家们的分歧和我们的困惑，希望总理帮助定夺。当时，温总理谈到抗震纪念园既要反映自然规律与灾害无情，更要表达中国人民抗震救灾与重建的伟大力量，还要展示新北川未来的美好前景。经过慎重思考，最终由中规院根据清华大学、中建院、北林地景和天津华汇等设计单位的方案，并结合多位院士和设计大师的建议，完成园区的总图方案。中规院城市设计研究室蒋朝晖先生以充满智慧的"俏皮"手法（吴良镛先生曾以此评价林璎的华盛顿越战纪念碑方案，我以为非常贴切），选取三轮方案中最有价值的理念和局部方案，"拼贴"出了三段式的总图方案。此后，宋春华理事长亲任指挥，张锦秋院士任顾问，周恺、孟建民、庄惟敏大师分别承担静思、英雄和幸福三个主题园的设计，叶毓山大师承担主题雕塑（纪念碑）设计，北林地景承担园林景观设计，中规院承担场地设计，共同完成了抗震纪念园的设计。

回望三年重建，应该铭记的建筑专家还有很多。何镜堂院士、马国馨院士、刘力大师、黄星元大师、朱自煊先生、李先逵先生、崔彤总建筑师和我院邹德慈院士、王瑞珠院士、王景慧老总等专家都作出了重要的贡献，付出了辛勤的劳动。北川新县城的成功在于规划师、建筑师与政治家，与公民对城市的风貌与品格达成了共识；在于规划师、建筑师从不同的方面，用不同的创意和手法服务于共同的目标；在于建筑师在设计过程中始终心中有一个"城市"的概念。我们可以自豪地说，北川新县城每一幢建筑都是建筑师尽心尽力设计的作品，都是城市整体中不可或缺的一个组成部分。这里没有鹤立鸡群、自高自大的"主角"，也没有潦草丑陋的"配角"。

北川重建是幸运的。四川、绵阳、北川和山东前指领导干部对专业工作深刻理解，对建筑创作充分尊重。绵阳市吴靖平书记多次参加重要建筑方案和纪念园主题雕塑等的评审与讨论。陈兴春书记、经大忠县长、王绵生指挥长、徐振溪指挥长等重建的直接领导，从未把任何个人观点强加给建筑师。多次到北川考察的党和国家领导、四川和山东两省领导也从未对具体建筑方案提出非专业的指示和要求。北川新县城正是得益于这种尊重专家、尊重创作的良好氛围，而没有出现一个"长官意志"的建筑方案。

非常感谢崔愷大师提议并主持编辑出版《建筑新北川》，感谢修龙院长和中建院为本书出版的投入，尤其是张广源摄影师，前后用了一年多的时间，跟踪北川重建的进展，捕捉最佳天气和光影，精心摄制了大量精美的照片。因为这些，才有了今天奉献给灾区民众、给规划、建筑同行的《建筑新北川》。我们仍在努力，争取尽早把《规划新北川》、《建设新北川》等奉献给大家。

2011年5月8日，温总理第十次到北川，还未等我开口，温总理就急切地问道："我们当时说好的六句话、十二个字，你们做得怎么样？"共和国总理对灾区重建规划设计的牵挂让我无比感动！在写这篇前言的时候，恰好有机会聆听吴良镛先生讲了三个发人深省的观点：第一，我们正处在一个伟大的时代，我们要无愧于这个时代；第二，中国优秀的人居文化传统与当今极为丰富的实践，比历史上任何理论要丰富和厚重得多；第三，我们应该创造适合中国、适合地方的全新发展模式。的确，中国正经历着前无古人的城市化与城市发展，刚刚完成了前无古人的汶川特大地震灾后重建。我想，政治家的治国方略能与规划师、建筑师的理想合拍，民众能通过规划师、建筑师的努力获得福祉，正是这个伟大时代给予规划师、建筑师的幸运！努力摆脱西方"经典"的束缚，从历史传统和民族文化中吸取养分，从本土与时代的双向纬度探索中国城市与中国建筑的现代表达，正是中国规划师、建筑师的文化自觉和历史责任。

感谢北川，感谢所有参加北川新县城重建的人们，让中国规划师与建筑师共同参与创造了一个伟大的奇迹。

Preface: A Model of Collaboration between Planners and Architects

Li Xiaojiang / President of China Academy of Urban Planning and Design
Commander in Chief of the Planning and Design for the New County Seat of Beichuan, China Academy of Urban Planning and Design

The reconstruction of Beichuan after the May 12th earthquake offers Chinese planners an opportunity to undertake social responsibilities and make contribution to the public with their professionalism, and also gives an opportunity to planners and architects in close collaboration and mutual communication.

On May 19, 2008, as team leader of Mianyang Task Force on Earthquake Emergency Planning organized by the State Ministry of Housing and Urban-Rural Development, I arrived at old county seat of Beichuan for the first time, and was deeply shocked by the ruthless disaster. When I went there for the second time on May 25, 2008, I was strongly impressed by Mr. Song Ming, secretary of Beichuan County and Mr. Jing Dazhong, head of the county, with their urgent desire of building a new homeland and their trust in planners. In the end of May 2008, *Preliminary Report on Beichuan Reconstruction Site Selection Demonstration* was finished in only 10 days jointly by China Academy of Urban Planning & Design (CAUPD) and Wuhan Institute of Geological Engineering Exploration. In early June 2008, this report was signed and issued by Mr. Tan Li, Mianyang municipal party secretary and submitted to the State Council Earthquake Relief Headquarters. On June 3, 2008, Vice Minister Qiu Baoxing of the State Ministry of Housing and Urban-Rural Development inspected the old county seat of Beichuan and the site selection for the new county seat, and instructed CAUPD to try every effort in post-quake reconstruction. In August 2008, Mr. Chen Xingchun, new secretary of Beichuan County party committee recommended that CAUPD be chief technical coordinator for the reconstruction. In October 2008, the planning for the new county seat of Beichuan was formally started. In the first half of 2009, guided by Mr. Wu Jingping, Mianyang municipal party secretary, a series of organizations were established including the Construction Headquarters for New Beichuan and the Front-Line Headquarters for New Beichuan Planning by CAUPD, forming a "three-headquarter coordination system" together with the Front-Line Headquarters of Aiding Projects to Beichuan by Shandong Province.

On November 16, 2008, Chinese Premier Wen Jiabao inspected the new county site, and pointed out that the new county seat of Beichuan should aim at becoming "a symbol for urban construction and engineering, a symbol of earthquake relief and a symbol of cultural heritage" under the principles of "safety, livability, characteristic, prosperity, advancement, harmony". On May 11, 2009, General Secretary Hu Jintao visited the new county seat and stressed that the reconstruction is anticipated by the local people of Beichuan and the whole country as well. Although CAUPD has undertaken numerous big planning projects over the past several decades, when facing such an extraordinarily important landmark project of building a new Beichuan, it really comes to realize the meaning of General Secretary Hu Jintao when he said it is a "heavy responsibility and glorious mission".

As Beichuan is the only autonomous county of the Qiang nationality, the biggest challenge for CAUPD as chief technical coordinator is how to pass on the Qiang historic culture, forms and styles through building its new county seat and architecture, how to make architectural design innovative while keeping overall harmony of the city, and how to complete several hundreds of architectural designs for projects totaling over 2 million km^2 in a short time.

Fortunately, the planning of the new county seat received warm response from many design masters and institutes, and Architectural Society of China (China ASC) gave its full support in particular. On June 28, 2009, "Expert Seminars on the Townscape and Building Style of New County Seat of Beichuan" were jointly held by China ASC, Mianyang Municipal Party Committee, Mianyang Municipal Government and CAUPD, attracting academicians, design masters, researchers and scholars of the western Sichuan and the Qiang folk houses, and local cultural and folk experts. At the seminars, *Six-point Consensus on Townscape and Building Style of New County Seat of Beichuan* was reached, which guides site selection and architectural design of new Beichuan.

Highly respected Song Chunhua, director-general of China ASC, and Zhou Chang, secretary-general of China ASC mobilized a host of famous scholars and masters to involve in the architectural scheme design, review and selection. They organized dozens of meetings in Beijing, Chengdu, Mianyang and Beichuan to strive for perfect work. China ASC has made irreplaceable contribution to the architectural design for new county seat of Beichuan.

Academician Zhang Jinqiu, at the key moments of choosing scheme for the Memorial Park, personally assumed the consultant and encouraged the planners to adjust and integrate schemes. In the spirit of perseverance, wise thinking and gentle attitude, he led the design team of CAUPD to constantly develop planning conception and modified the scheme personally. It is the utmost guidance of academician Zhang Jinqiu that a work of collateral wisdom is generated, which represents the most important spiritual place in the new Beichuan.

Master Cui Kai won unanimous praise for his design of the Cultural Center. With a strong background in study on local building traditions and the Qiang courtyards, he creates a fresh new "Qiang village" from the exterior to the interior by utilizing local practice and materials, and is constantly persisting in his design concept. Now, a modern Qiang "village" stands at the east end of the cultural axis in the new county seat of Beichuan.

The Happiness Pavilion is a 2,000m^2 building designed by Master Zhuang Weimin. During scheme modification, Master Zhuang Weimin adhered to a "white stone" concept and he also modestly coordinated with the general planning, showing distinctive features of the design in harmony with the entire background.

The Meditation Garden is a quite emotional venue in the new county seat designed by Master Zhou Kai who experienced and survived in the Tangshan earthquake in 1976 and who produces such a touching design with his unique perception of life. The Meditation Garden is a place to have a mild dialogue with the dead and the heaven, where people can meditate on the deep meaning of life in silence, and ponder on the natural force in plainness.

I particularly admire Mater Meng Jianmin's dedication to design the Hero Garden in Memorial Park who insisted on designing petty pieces in a large background by deliberately designing ground pavement and lamps with a visual effect of extensive curve.

Mr. Sheng Ye and Mr. Lu Qiang, general and vice general managers of Hong Kong Huayi Architectural Design Co., Ltd are enthusiastic in design of a new Beichuan. Their active attitude and corporate strength make the company the only one which wins the total three architectural design competitions. The scheme of the Administrative Center designed by Chen Ribiao is solemn and grand with no exaggeration, which pursues organic inheritance and innovation of the Qiang architectural culture, and strives for a perfect combination of scale, size, material and color.

Prof. Deng Xuexian from Tsinghua University and Prof. Ji Fuzheng from Southwest Jiaotong University completed the detailed planning and architectural scheme for the Qiang Features Pedestrian Street and Yuwang Bridge in spite of illness. With strong traditional Qiang styles, these two places become tourism highlights in the new county seat.

Recommended by Director General Song Chunhua of China ASC, renowned sculptor Ye Yushan designed the theme sculpture for the Memorial Park—*Rebirth*. This piece of work not only demonstrates the heroic spirit of the Chinese people in conquering difficulties, but also showcases vision of the Qiang people for good life. This masterpiece is a core visual focus in the new county seat, showing rebirth of a nation and a town. As planner of new Beichuan, I especially appreciate the work *Rebirth* as a finishing touch for the new county.

China Architecture Design & Research Group (CAG) is the strongest partner of the architectural design for the new county seat. President Xiu Long of CAG stressed on many occasions that CAG be a good technical supporter of CAUPD. The team led by Liu Yanhui, secretary of the Party committee of CAG, is the first to cooperate with the architectural design team of CAUPD, and has completed the schematic design, preliminary design and landscape design for all the 1.2 million m^2 residential buildings of the new county seat, the resettlement housing thus become the earliest started and completed project, honoring the commitment of giving priority to people's well being in post-quake reconstruction. In order to ensure that all residential buildings meet the "green building" standard, Secretary Liu guided to find ways of saving construction cost. In the late phase of reconstruction, CAD undertook a series of architectural designs for office and service buildings which seemed "small" but closely related to full-fledged functions of the new county seat. The landscape design for the Er'ma Residential Quarter made by CAD integrates many Qiang cultural elements, and is welcomed by the quake-affiliated people; the flexible layout of the Yongchang Primary School, its spatial organization of courtyard, and beautiful and elegant appearance are innovative explorations on school building design, which are highly appreciated by the teachers and students of the school.

The Architectural Design Department of CAUPD (also named Beijing Guocheng Architectural Design Company), although not as competent as large design institutes, is always cooperative and modest in working with CAG. It has completed all the preliminary designs for residential buildings by partnering with CAG, and undertaken a huge amount of construction drawing designs and site designs for residential buildings and designs for a batch of public buildings. The department/company assumed the largest workload of architectural design which stayed on the site for the longest time for supporting planning and construction.

The reconstruction of the new county seat of Beichuan is a model of mutual understanding, support and close collaboration among politicians, planners and architects, of which, the Memorial Park is a case in point. The three rounds of scheme recruitment organized by ASC generated more than twenty schemes, and many of them are excellent works of academicians and masters. However, no consensus was reached on the theme expression of the Memorial Park in the three rounds. After discussing with director-general Song Chunhua, we decided to directly report to Premier Wen Jiabao for his instruction when he paid a site visit to the new county. On September 25, 2009, I made a report on the planning and design for the new county seat to Premier Wen Jiabao, and the experts' different understandings as well. Premier Wen noted then that, in the meantime of reflecting the merciless laws of nature and disaster, the Memorial Park design should also demonstrate great power of the disaster relief and reconstruction of the Chinese people, and the bright future of new Beichuan. Through careful consideration, the master plan scheme for the park is finally completed by CAUPD by referring the design schemes made by Tsinghua University, CAG, Beijing Beilin Landscape Architecture Institute Co., Ltd. and Tianjin Huahui Architectural Design & Engineering Co., Ltd and incorporating the suggestions of academicians and masters. Mr. Jiang Zhaohui from the Urban Design Department of CAUPD, in a wise manner, selected the most valuable essence from the three rounds, and collaged a master plan scheme. Based on this, the design for Memorial Park was complete with director-general Song Chunhua of ASC as commander, academician Zhang Jinqiu as consultant, master Zhou Kai, Meng Jianmin and Zhuang Weimin as designers for the Meditation Garden, the Hero Garden and the Happiness Pavilion respectively, master Ye Yushan as sculptor for theme sculpture, and Beijing Beilin Landscape Architecture Institute Co., Ltd. and CAUPD for the landscape design and site design respectively.

In retrospect of the past three years of building the new county seat of Beichuan, we still have a lot of building experts to be memorized, for instance, Academician He Jingtang and Ma Guoxin, master Liu Li and Huang Xingyuan, Mr. Zhu Zixuan, Shan Deqi, Li Xiankui and Cui Tong, chief architect, academician Zou Deci and Wang Ruizhu from CAUPD and Wang Jinghui, chief planner of CAUPD, they make great contribution to the reconstruction of the new county seat of Beichuan. The planners, architects, politicians and general public reach consensus on the style and characteristics of the new county seat of Beichuan; the planners and architects, with their creations, adopt different methods to realize a common goal from different perspectives; and the architects always have a "city" concept in their minds during the design process. So we can say with proud that each piece of design is made dedicatedly for the utmost interest of the public, properly playing their respective roles.

Fortunately, in the whole process, leaders of Sichuan province, Mianyang city, Beichuan County and the Front-Line Headquarters of Aiding Projects to Beichuan by Shandong Province showed their greatest understanding and respect on the professional works. Mr. Wu Jingping, Mianyang municipal party secretary, personally participated in reviews and discussions on major architectural schemes and theme sculpture of the Memorial Park. Major leaders including Secretary Chen Xingchun, Secretary of Beichuan County Party Committee, Jing Dazhong, head of Beichuan County, Commander Wang Jinsheng and Xu Zhenxi never impose their personal opinions on architects. Leaders of CPC and the state visited Beichuan for many times, and leaders of Sichuan and Shandong provinces never put forward non-professional instructions and requirements on the specific architectural schemes. Thanks to this mutual understanding and cooperation, the reconstruction of new Beichuan can be progressed smoothly with no "will of leading officials".

I am especially grateful for Master Cui Kai who proposed and organized editing and publishing Building a New Beichuan, President Xiu Long and CAG for their efforts in publishing this book. In particular, I'd like to thank Mr. Zhang Guangyuan, renowned architectural photographer, who took more than a year tracking the development of reconstruction of the new county seat of Beichuan and captured a huge amount of glorious photos. Their joint efforts make it possible to present *Building A New Beichuan* to the quake-affiliated people, planners and architects. And we are working to present readers *Planning A New Beichuan* and *Constructing A New Beichuan* at the earliest date.

On May 8, 2011, Premier Wen Jiabao visited Beichuan for the tenth time since the earthquake. Before I started to report to him, Premier Wen asked me eagerly about the implementation of the "Six Synergies" and "Twelve Principles" (in Chinese characters) for planning and design. On this occasion of writing the preface, I am lucky to talk with Mr. Wu Liangyong. Mr. Wu proposed three thought-provoking ideas: firstly, living in a great era, we should be worthy of the great time. Secondly, the excellent habitat culture tradition of China and rich practices at present are abundant than any theories in the history. Thirdly, we should create a new developing mode that fits China and local places. Indeed, China is now experiencing unprecedented urbanization and urban development, and the post-quake reconstruction of Beichuan after the May 12th Earthquake is just completed. I believe that planners and architects in this context are lucky as their ideas are consistent with the governing strategies of the politicians, and the general public can get benefits from their efforts! It is cultural awareness and historical responsibility of planners and architects to free themselves from the so-called "classic" of the western countries, absorb nutrients from our long history and national features, and explore modern expression of Chinese cities and architecture from both the perspectives of modernization and localization.

We should thank Beichuan and all those who involved in the reconstruction of the new county seat of Beichuan for making Chinese planners and architects jointly accomplish a great historical mission!

北川新县城灾后重建规划与建筑设计工作大事记

Milestone of Post-quake Reconstruction Planning and Architectural Design for the New County Seat of Beichuan

2008年

5月12日，我国发生汶川特大地震，伤亡惨重、损失巨大。

5月19日，中国城市规划设计研究院（中规院）汶川地震抗震救灾绵阳工作组启动北川新县城选址工作。

6月3日，中规院汶川地震抗震救灾绵阳工作组提出北川新县城选址方案。

8月，四川省政府正式向国务院报送了北川新县城选址意见。

11月7日，北川新县城灾后重建总体规划工作动员大会召开。

11月10日，国务院常务会议原则通过北川新县城选址。

11月16日，温家宝总理实地考察北川新县城选址，提出北川新县城规划设计要以"安全、宜居、特色、繁荣、文明、和谐"为标准，以把北川新县城建设成为"城建工程标志、抗震精神标志和文化遗产标志"为目标。

12月20日，住房和城乡建设部与四川省建设厅联合组织"北川新县城灾后恢复重建总体规划技术审查"。

2009年

2月6日，民政部正式批复北川羌族自治县行政区划调整方案，将包括北川新县城选址在内的208km²从安县划入北川县行政区划范围。

2月9日，北川县举行元宵晚会，中规院向到会受灾群众详细讲解"北川新县城灾后重建总体规划"方案。

3月18日，北川县灾后重建城乡规划委员会正式成立并举行第一次会议，审议通过了新县城红旗拆迁安置区修建性详细规划和温泉片区受灾群众安居房修建性详细规划方案。

3月30日，四川省政府正式批复《北川羌族自治县灾后恢复重建总体规

2008

May 12th An Earthquake hit Wenchuan, Sichuan Province of China, causing great casualties and huge losses.

May 19th Mianyang Earthquake Relief Task Force of China Academy of Urban Planning & Design (CAUPD) launched site selection for the new county seat of Beichuan.

June 3rd CAUPD Post-quake Reconstruction and Planning Task Force proposed site selection scheme for the new county seat of Beichuan.

August Sichuan Provincial People's Government submitted comments on site selection of the new county seat of Beichuan to the State Council.

November 7th Mobilization Meeting on Comprehensive Plan for Beichuan Qiang Autonomous County was held.

November 10th The site selection scheme for the new county seat of Beichuan was approved in principle at the executive meeting of the State Council.

November 16th Chinese Premier Wen Jiabao paid a site visit to the proposed site of the new county seat of Beichuan, and pointed out that planning and design for the new county seat of Beichuan should adhere to the principle of building a "safe, comfortable, vigorous, featured, civilized and harmonious" town, and the goal is to build the new county seat of Beichuan into "a symbol of urban construction, a symbol of quake relief spirit and a symbol of cultural heritage".

December 20th The Ministry of Housing and Urban-Rural Development and Department of Construction of Sichuan Province jointly organized "Technical Review on General Planning of Post-quake Reconstruction of Beichuan Qiang Autonomous County."

2009

February 6th The Ministry of Civil Affairs of the People's Republic of China officially approved the adjustment on administrative district scheme of Beichuan Qiang Autonomous County, incorporating an area of 208 km² (including the site selected for the new county seat of Beichuan) formerly under the jurisdiction of Anxian County into the administrative district scope of Beichuan County.

February 9th At the Lantern Festival party held in Beichuan, CAUPD explained in detail the scheme of "General Planning of Post-quake Reconstruction of the new county seat of Beichuan" to earthquake-affiliated people present at the party.

March 18th Beichuan County Post-quake Reconstruction Planning Committee was officially established. At the first

2009年5月11日 胡锦涛总书记视察北川新县城工地

General Secretary Hu Jintao visited the construction site of the new county seat of Beichuan on May 11th, 2009.

（摘自《北川新县城建设志》，*The Construction Chorography of the New County Seat of Beichuan*）

2008年11月16日 温家宝总理实地考察北川新县城选址

Premier Wen Jiabao paid a site visit for selecting location for the new county seat of Beichuan on November 16th, 2008.

（《绵阳日报》蒲涛摄，Photo by Pu Tao, *Mianyang Daily*）

划》。

4月22日，北川县灾后重建城乡规划委员会举行第二次会议，审议通过了红旗片区拆迁安置房和温泉片区受灾群众安居房建筑设计方案，研究讨论新县城住房政策。

4月30日，北川县灾后重建城乡规划委员会举行第三次会议，审议通过了新县城园林绿地系统规划。

5月10日，中规院北川新县城规划工作前线指挥部成立。

5月11日，胡锦涛总书记视察北川新县城工地，听取中规院关于北川新县城规划工作汇报，接见中规院等现场规划设计专家和山东援建工程技术人员。

5月12日，新北川中学举行开工奠基仪式。

5月30日，北川新县城和北川山东工业园规划建设指挥部成立。

6月8日，山东省援建北川新县城暨山东产业园第一批项目开工仪式，标志着新县城建设全面开工。

6月16日，北川新县城建筑设计方案第一次专家评审会召开，分别选定永昌小学、永昌中学、计生指导站、爱卫办、就业社会保障综合服务中心，敬老院、福利院等项目推荐方案。

6月22至23日，北川新县城第一批建筑设计方案第二次专家评审会召开，就新北川宾馆、惠民大楼等10处建筑设计项目共40个方案进行

meeting of the committee, participants reviewed and agreed the constructive detailed plan of Hongqi resettlement area and Wenquan resettlement area.

March 30th Sichuan Provincial People's Government officially approved *Planning of Post-quake Reconstruction of Beichuan Qiang Autonomous County.*

April 22nd Beichuan County Post-quake Reconstruction Planning Committee held the second meeting, at which participants reviewed and agreed the architectural design schemes for both Hongqi resettlement area and Wenquan resettlement area, and housing policy of the new county seat of Beichuan was also discussed at the meeting.

April 30th At the third meeting of Beichuan County Post-quake Reconstruction Planning Committee, participants reviewed and agreed the Plan of Green Space of new county seat of Beichuan.

May 10th The CAUPD Planning Front-Line Headquarters for the new county seat of Beichuan was established.

May 11th General Secretary Hu Jintao inspected the construction site of the new county seat of Beichuan, listened to the report on the planning and met with on-site planners and designers from CAUPD and technical personnel involved in the projects aided by Shangdong Province.

May 12th The foundation laying ceremony of new Beichuan High School was held.

May 30th Planning & Construction Headquarters for the new county seat of Beichuan Beichuan-Shandong Industrial Park were established.

June 8th The commencement ceremony of the first batch of projects aided by Shandong Province and Shandong Industrial Park was held, which is a full start of building the new county seat of Beichuan.

June 16th The first expert review meeting on the architectural design scheme of the new county seat of Beichuan was held, and the schemes reviewed at the meeting include projects of Yongchang Primary School, Yongchang Middle School, Family Planning Service Building, Office of Patriotic Health Campaign Committee, Employment and Social Security Comprehensive Service Center, Senior Citizens' Home and Welfare House, etc.

2008年 中规院工作人员在进行新县城选址踏勘、调研

CAUPD staff made site survey and investigation for selecting location for the new county seat of Beichuan in 2008.

2009年2月9日 中规院在元宵晚会上讲解新县城规划方案

At the Lantern Festival party held in Beichuan, CAUPD explained the scheme of new county seat general planning on February 9th, 2009.

2009年3月18日 北川县灾后重建城乡规划委员会第一次会议

The kickoff meeting of Beichuan New County Seat Post-quake Reconstruction Planning Committee on March 18th, 2009.

（中规院供稿，by CAUPD）

2009年9月25日 温家宝总理视察指导北川县灾后恢复重建工作，接见中规院规划专家

Premier Wen Jiabao inspected the earthquake rehabilitation and reconstruction work of Beichuan County and met planning experts of CAUPD on September 25th, 2009.

2009年2月22日 中建院和中规院专家考察安居房规划用地

The experts of CAG and CAUPD surveyed the affordable housing site on February 22nd, 2009.

（中规院供稿，by CAUPD）

2009年5月12日 新北川中学举行开工奠基仪式

The foundation laying ceremony of new Beichuan High School on May 12th, 2009.

（摘自《北川新县城建设志》，*The Construction Chorography of the New County Seat of Beichuan*）

评审。

6月29日，北川新县城第一批建筑设计方案第三次专家评审会召开，就羌族特色商业步行街与禹王桥、体育中心两处建筑设计项目共10个方案进行评审。

7月13至14日，中国建筑学会、中国城市规划学会城市设计学术委员会、中规院在绵阳召开"北川新县城城镇风貌与建筑风格专家研讨会"，确定北川新县城城镇风貌与建筑风格既要继承传统民族建筑文化体现羌族风韵，又要体现现代元素，用于指导新县城建筑设计方案的"六点共识"。

7月23日，北川县灾后重建城乡规划委员会举行第四次会议，审议通过了新县城城镇风貌及建筑风格指导意见和新县城第一批建筑项目设计征集方案。

7月31日，北川县灾后重建城乡规划委员会举行第五次会议，审议通过了新县城重要公共建筑项目的建筑设计方案，研究讨论了安昌河拦蓄水工程及景观设计方案。

8月28日，北川县灾后重建城乡规划委员会举行第六次会议，原则通过了北川新县城部分公共建筑项目的初步设计、北川——山东产业园的详细规划方案、新县城公共服务设施项目梳理方案和安昌镇总体规划。

9月24日，北川县灾后重建城乡规划委员会举行第七次会议，审议通过了

June 22nd~23rd The second expert review meeting on the first batch of architectural design schemes of the new county seat of Beichuan was held, and 40 schemes were reviewed at the meeting, including architectural design for projects of New Beichuan Hotel, Livelihood Building, etc.

June 29th The third expert review meeting on the first batch of architectural design schemes of the new county seat of Beichuan was held, and 10 schemes were reviewed at the meeting, including architectural design for projects of the Qiang Features Pedestrian Street, Yuwang Bridge and Sports Center.

July 13th~14th Architectural Society of China, Acadmic Committee of Urban Design of Urban Planning Society of China and CAUPD jointly held "Expert Seminars on the Townscape and Building Style of the new county seat of Beichuan" in Mianyang. The "six-point consensus" was reached at the meeting which can guide the architectural schematic design for the new county seat of Beichuan, and the essence of the "six-point consensus" lies in that townscape and architectural style of the new county seat of Beichuan should be a combination of traditional national architectural culture reflecting the Qiang features and modern elements.

July 23rd The fourth meeting of Beichuan County Post-quake Reconstruction Planning Committee was held, and participants reviewed and agreed the instructional comments on the townscape and building style of the new county seat of Beichuan as well as schemes collected for the first batch of construction projects of the new town.

July 31st The fifth meeting of Beichuan County Post-quake Reconstruction Planning Committee was held, and participants reviewed and agreed the architectural design schemes for important public building projects, and they also discussed Anchang River Impounded Water Project and landscape design scheme.

August 28th The sixth meeting of Beichuan County Post-quake Reconstruction Planning Committee was held, and companies presented at the meeting passed in principle the preliminary designs for some public building projects, detailed planning of Beichuan-Shandong Industrial Park, scheme for public service facilities and general planning of Anchang Town.

September 24th The seventh meeting of Beichuan County Post-quake Reconstruction Planning Committee was held, and companies presented at the meeting agreed the constructive detailed plan of resettlement residential quarter in Baiyangping Area and Phase II of Wenquan resettlements area and passed in principle the design schemes for several

从云盘山看新县城施工进程(上 2010年1月　中 2010年6月　下 2010年10月)(中规院供稿)

The construction process, viewing from Yunpan Mountain. (above January, 2010 middle May, 2010 below October, 2010) (by CAUPD)

拆迁安置小区修建性详细规划和受灾群众安居房二期修建性详细规划方案等，原则通过了安昌河数座桥梁的设计方案。

9月25日，温家宝总理视察指导北川县灾后恢复重建工作，接见中规院规划专家，提出要进一步做好规划设计专家论证和公众参与工作，强调"以人为本、科学重建"是重建工作的灵魂，并对抗震纪念园的规划构思提出明确建议。

10月27至28日，建设部与四川省政府在绵阳联合召开北川新县城规划建设工作推进协调会，进一步提升北川新县城规划设计成果，建立新县城建设部、省、市、县联动机制。

12月8日，行政事业单位办公楼建筑方案专家评审。

12月29日，永昌大道两侧央企办公楼建筑设计方案专家咨询会。

2010年

1月4日，市县联合组建的北川新县城规划委员会正式成立并举行第一次会议，审议并原则通过了《抗震纪念园设计方案》、《北川新县城路灯设计方案》、《北川新县城交通管理设施设计方案》等六个方案。

1月18日，中规院组织抗震纪念园深化方案专家讨论会，中国建筑学会理事长宋春华、工程院院士张锦秋等参加。

3月21日，北川新县城规划委员会第二次会议召开，审查通过了抗震纪念

bridges over the Anchang River.

September 25th Premier Wen Jiabao inspected rehabilitation and reconstruction work and met planning experts of CAUPD; he instructed to proceed with the planning expert panel discussion and public participation, stressed that "human-oriented and scientific reconstruction" is the essence of the reconstruction of the new county seat of Beichuan, and proposed definite suggestions on the planning concept of the Memorial Park.

October 27th~28th The Ministry of Construction of China and Sichuan Provincial People's Government jointly held Promotion and Coordinating Meeting on the Planning and Construction of the new county seat of Beichuan in Mianyang to further consolidate the achievements of the new county seat of Beichuan planning and design, and establish a linkage system integrating the ministerial, provincial, municipal and county-level organizations.

December 8th Expert review meeting on the architectural scheme of office building for administrative institutions was held.

December 29th Expert consultation meeting on the architectural design scheme of office building for central enterprises on the both sides of the Yongchang Avenue was held.

2010

January 4th The Planning Committee of Beichuan New County jointly established by the municipal and county-level organizations was officially founded and the first meeting of the committee was held, at which participants reviewed and passed in principle the six schemes including *Design Scheme of Memorial Park, Road Lamp Design Scheme and Design Scheme of Traffic Control Facilities,* etc.

January 18th CAUPD organized Expert Discussion Meeting on the Design Development Scheme of Memorial Park, and Song Chunhua, director-general of Architectural Society of China and Zhang Jinqiu, academician of Chinese Academy of Engineering attended the meeting.

March 21st The second meeting of the Planning Committee of Beichuan County was held, at which participants reviewed and passed design scheme for theme sculpture of "Rebirth" in Memorial Park, the planning and design scheme for

山东省援建北川新县城工地现场（杜洪岭供稿）

Construction site of the project aided by Shandong Province. (by Du Hongling)

2010年11月3日 欢送山东援建人员仪式（中规院供稿）

Farewell ceremony for Shandong Province Aiding Crew on November 3rd, 2010. (by CAUPD)

2010年1月31日 永昌镇政府正式成立（中规院供稿）

The Government of Yongchang Town was officially founded on January 31st, 2010. (by CAUPD)

2010年9月25日 山东省对口援建北川新县城项目竣工交接（贺旺供稿）

Hand-over ceremony for the completion of the project aided by Shandong Province on September 25th, 2010. (by He Wang)

2010年11月3日 欢送山东援建人员仪式（孙彤供稿）

Farewell ceremony for Shandong Province Aiding Crew on November 3rd, 2010. (by Sun Tong)

园主题雕塑《新生》方案、北川新县城抗震纪念园规划设计方案和绿化景观提升方案。

4月19日，北川新县城规划委员会研究审议《北川地震纪念馆建筑设计方案》、《关于北川新县城抗震纪念园主题雕塑完善方案及实施办法》等。

9月1日，新北川中学举行〝重生〞后的第一个开学典礼。

9月19日，北川——山东产业园12家入园企业竣工投产。

9月25日，山东省对口援建北川项目竣工交接仪式在新县城举行，援建双方签署《山东省——北川羌族自治县合作协议》。

12月18日，首批入住北川新县城拆迁安置房摇号分配仪式举行。

12月23日，老县城建成区受灾群众首批入住新县城。

Memorial Park and the scheme for greening landscape improvement.

April 19th The Planning Committee of Beichuan New County discussed and reviewed *Architectural Design Scheme for Beichuan Earthquake Memorial* and *Improvement Scheme for theme sculpture in Memorial Park and Implementation Methods* and other schemes.

September 1st The first opening ceremony of new Beichuan High School after "rehabilitation" was held.

September 19th Twelve enterprises in the Beichuan-Shandong Industrial Park completed and put into operation.

September 25th The hand-over ceremony for the completion of the project aided by Shandong Province was held in the new county seat of Beichuan, and the two sides signed *Cooperation Agreement between Shandong Province and Beichuan Qiang Autonomous County.*

December 18th The first batch of residents of the resettlement housing in the new county seat of Beichuan was selected through computer selection and distribution.

December 23rd The first batch of residents of the resettlement housing in the new county seat of Beichuan who once lived in the built-up area of Beichuan was selected through computer selection and distribution.

2011年

2月1日，虎年腊月廿九，以〝开启永昌之门 点燃幸福之火〞为主题的开城仪式举行，标志着北川新县城正式开城。

4月18日，四川省旅游发展大会北川新县城分会场开幕，羌族特色步行街开街。

5月8日，温家宝总理第十次来到北川，视察北川中学、羌族特色步行街和安置小区，接见规划设计人员，充分肯定新县城规划设计工作。

2011

February 1st On the 29th day in the last month of the lunar calendar of the Year of the Tiger, a Town Opening Ceremony themed by "opening gate of prosperity• lighting flame of happiness" was held, marking the official opening of the new county seat of Beichuan.

April 18th The parallel session of Sichuan Provincial Tourism Development Conference inaugurated and Qiang Feature Pedestrian Street opened to the public.

May 8th Premier Wen Jiabao went to Beichuan for the tenth time since the 5.12 Wenchuan Earthquake. He inspected Beichuan High school, Qiang Feature Pedestrian Street and resettlement quarters, met with planners and designers and fully affirmed the planning and design of the new county seat of Beichuan.

2010年12月18日 首批入住北川新县城拆迁安置房摇号分配仪式举行（贺旺摄）

The first batch of residents of the resettlement in the new county seat of Beichuan was selected through draw lots selection on December 18th, 2010. (by He Wang)

2010年12月23日 老县城建成区受灾群众首批入住新县城（欧阳均军摄）

The first batch of residents who once lived in the old county seat of Beichuan moved into the resettlement of the new county seat on December 23rd, 2010. (by Ouyang Junjun)

2011年2月1日　北川新县城举行开城仪式（左一 孙彤供稿　右二 县委宣传部供稿）

Town Opening Ceremony was held in the new county seat of Beichuan on February 1st, 2011. (by Sun Tong and Propaganda Department of County Party Committee)

2011年4月18日　羌族特色步行街开街（左上 县委宣传部供稿　左下、中下 欧阳均军供稿　中上、右一孙彤供稿 ）

The Qiang Features Pedestrian Street opened to the public on April 18th, 2011. (By Ouyang Junjun, Sun Tong and Propaganda Department of County Party Committee)

再造一个新北川 ——北川新县城规划设计及城镇风貌规划

朱子瑜 / 北川新县城规划设计项目负责人、中规院北川新县城规划工作前线指挥部执行副指挥

李　明 / 中规院北川新县城总体城市设计项目负责人

贺　旺 / 北川新县城建设指挥部副指挥，绵阳市规划局副局长

孙　彤 / 北川新县城控制性详细规划项目负责人、中规院北川新县城规划工作前线指挥部总指挥助理

北川是我国唯一的羌族自治县，位于山峦起伏的绵阳市西部。县城曲山镇周边龙门山数峰夹峙、城中湔江一水湍流。县城规模不大、名气不响、经济也不发达，但这里风景优美、生活安宁、社会和谐、羌族风貌浓郁。然而，小城却坐落在北川——映秀和擂鼓两大地震断裂带的交汇处，一个巨大的危机就潜伏在这平静的生活之下。2008年5月12日14时28分，中国西南大地轰然一震，天降劫难，曲山镇这座曾经秀丽的小城刹那间被夷为平地。山体滑坡、房屋倾倒、桥梁断裂、道路塌陷，曲山镇成为此次汶川特大地震的极重灾区（图1）。

图1 北川老县城震后景象

Fig. 1 Old county seat of Beichuan after the earthquake

曲山镇地处地质灾害风险极高的地区。受周边山体限制，平坝用地相当局促，原址重建既不合适、也无可能。当地受灾群众对未来县城迁往更安全、更平坦、更方便的新址也寄予厚望。2008年5月19日，绵阳市委、市政府请求国家住房和城乡建设部派专家为北川新县城规划选址，受住房和城乡建设部指派，中国城市规划设计研究院的专家团队到达北川，启动北川新县城选址工作，并由中国城市规划设计研究院、武汉勘测设计研究所、绵阳市规划局、中共北川县委、县人民政府共同编制完成了《北川县城"5•12"特大地震灾后重建选址与规划研究》。研究报告根据地质、交通、空间发展条件、区域城镇化发展要求和行政区划调整的可能，及时对若干选址进行了比较研究，后经中央批准，举世瞩目的北川新县城将选址于安昌镇东南方向，面积约10km^{2}的河谷平坝至盆地的过渡地段（图2）。

北川新县城是5・12汶川特大地震灾难后中央决定的唯一一个异地重建的县城。北川新县城的规划建设肩负着展示灾后重建成就和抗震精神、展示民族风貌和地域特色的历史使命。中国城市规划设计研究院自2008年5月14日第一时间深入北川救灾现场工作，先后开展了北川新县城的选址、总体规划、城市设计和专项规划等工作，并会同50余家规划设计单位，完成了市政道路工程设计、公共建筑方案设计、安居房建筑方案设计、绿地景观设计等一系列规划设计工作，努力以技术手段诠释北川新县城建设"安全、宜居、繁荣、特色、文明、

图2 新县城选址研究

Fig. 2 Relocation research of new county seat

图3 功能分区

Fig. 3 Function functional loyout

和谐”的十二字方针和“城建工程标志、抗震精神标志、文化遗产标志”三个标志要求。

一、北川新县城总体规划概述

根据党中央“再造一个新北川”的号召，全面贯彻国务院“十二字方针”和“三个标志”的规划工作要求，以“政府主导、专家领衔、部门合作、公众参与、科学决策”为工作原则，以妥善安置灾民、恢复县城功能、为城市长远发展奠定空间框架为核心任务，以先进理念与实施措施结合、规划方案与项目建设结合、专项规划与工程建设结合、规划控制与建设管理结合、政府决策与民众意愿结合、规划布局与城市设计结合为工作方法，以多专业协作、多方面互动为工作机制，又快又好地推进重建落实与实施。

北川新县城总体规划明确未来北川新县城将融入绵（阳）、江（油）、安（县）一体化发展格局。至2020年北川新县城将建成北川县域政治、经济和文化中心；川西旅游服务基地和绵阳西部产业基地；现代化的羌族文化城和生态园林城。近期用地规模5km²，人口规模3.5万；远期用地规模7km²，人口规模7万。根据城市发展目标和现状自然条件的特点，未来北川新县城将由南部山东产业园区、中部平坝综合生活区和北部丘陵旅游休闲度假区和西部未来功能拓展区等四大功能片区组成。南部山东产业园区以产业发展、解决居民就业为主要功能。产业园区内的产业门类以符合当地资源条件、保护环境为先、延续原有工业基础为前提，提供大约1万个就业岗位。园区提供完善的配套设施，并设置职业学校，使当地居民尽快获得就业技能和岗位。中部平坝综合生活区，地势平坦、水系发达，适合安排城市的生活及公共服务功能。至2010年将集中建设9000套左右保障性住房，满足曲山镇受灾居民、部分失地农民和当地拆迁村民的安置需求；至2015年将继续建设2600套左右保障性住房，规划期末保障性住房的建筑面积将占到整个住房建筑面积的46%。北部丘陵旅游休闲度假区，地形复杂、用地零碎、环境优良、生态敏感，规划以旅游度假休闲、文化产业和福利康复等功能为主（图3）。

北川新县城总体规划提出“山水环、生态廊、休闲带、生长脊、设施链、景观轴”的空间结构设计构思，以环、廊、带、脊、轴、链等多空间要素的相互交织构成了新县城功能布局与风貌展示的空间骨架，为快速重建背景下的城市发展提供清晰、有序、高效的城市空间格局。在山环水绕的山水格局下，安昌河作为新县城贯穿南北的“生态廊”，河道两岸绿化空间作为改善县城环境、保证静风频率较高地区小气候质量的重要生态走廊，也是重要的城市休闲健身活动空间。城市主要公共建筑沿道路骨架，形成城市公共服务“设施链”，集中安排城市主要公共服务设施，包括文化中心、抗震纪念园、惠民大楼（行政服务中心）以及学校、医院和体育中心等。以永昌河为主体的“休闲带”，结合原有水系梳理设置由一系列特色公园组成的城市公园带，将历史古迹保护、青少年和老年户外活动、抗震纪念、体育休闲等内容与滨河公园有机地组织在一起。巧妙借用周边自然山形和河道走势建立横贯东西的城市“景观轴”，有序统领整个城市的功能和空间景观。以周围山体为端景，景观轴既贯通城市的山水空间又串联文化中心、抗震纪念园、羌族特色步行街和步行风雨桥等实体功能，是城市重要的步行廊道，虚实结合成为北川新县城空间结构和功能结构的中枢。（图4）

二、北川新县城总体城市设计概述

北川新县城作为国内唯一羌族自治县的全新城镇，肩负地域民族文化的传承与复兴的光荣使命。在千百年的本族文化和外来文化的交织融合中，羌族传统聚落形成了因地借势，精于水工，有机自然的空间营造精髓，空间布局与形式不拘于定式定法的同时依靠相同材质与工艺的模仿建造行为形成了整体统一、环境协调的聚落整体风貌。而新县城风貌特色的建构与营造面临的是从传统到现实，从村寨到城市的聚落地理环境变迁、规模尺度扩大、功能组织复杂、民众生活方式由内向防卫转向社会交往的外向化、建

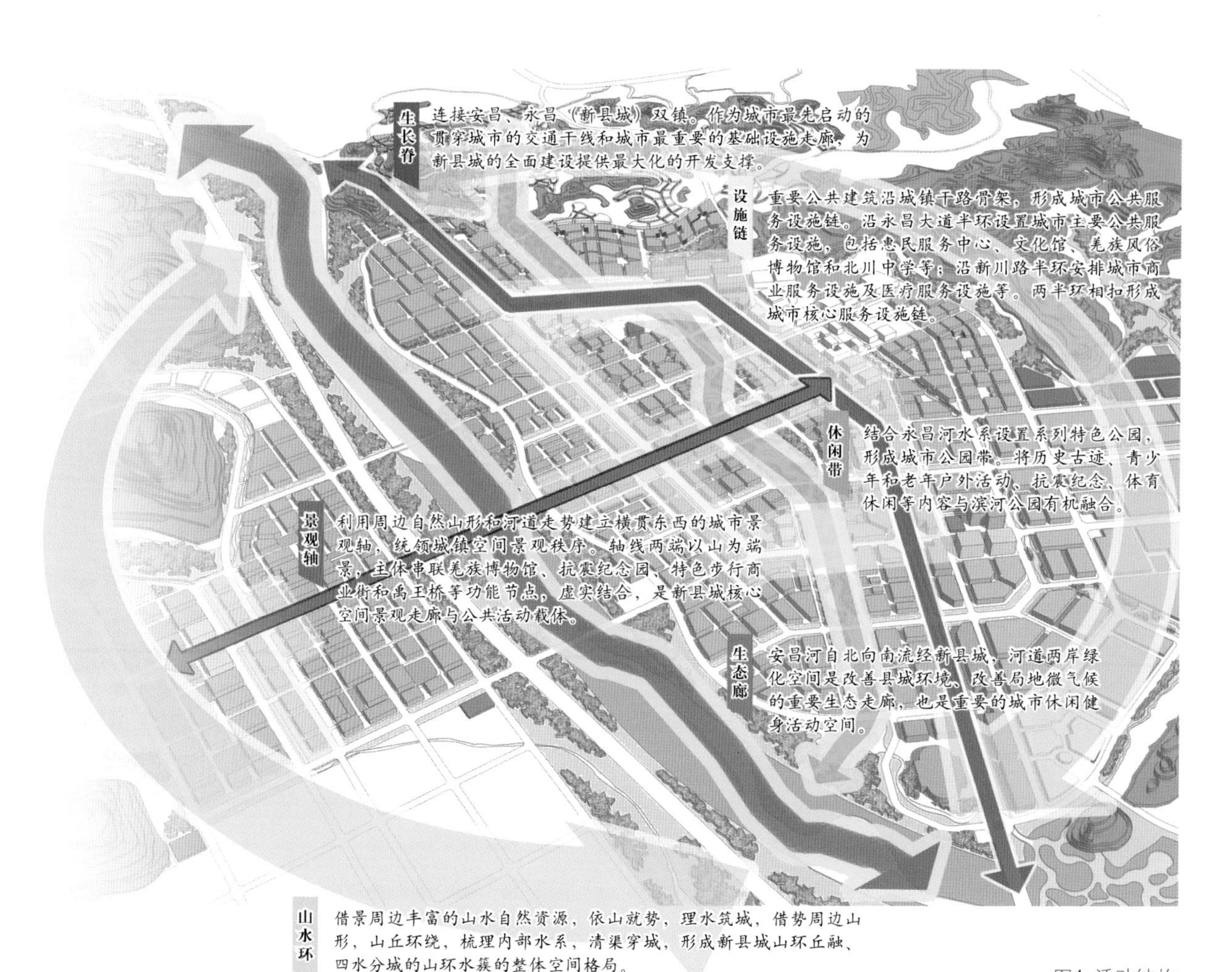

图4 活动结构

Fig. 4 Activity Structure

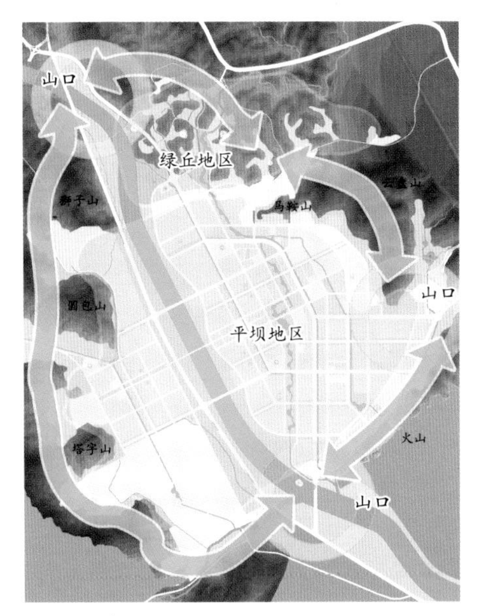

图5 山体格局 Fig. 5 Hill background

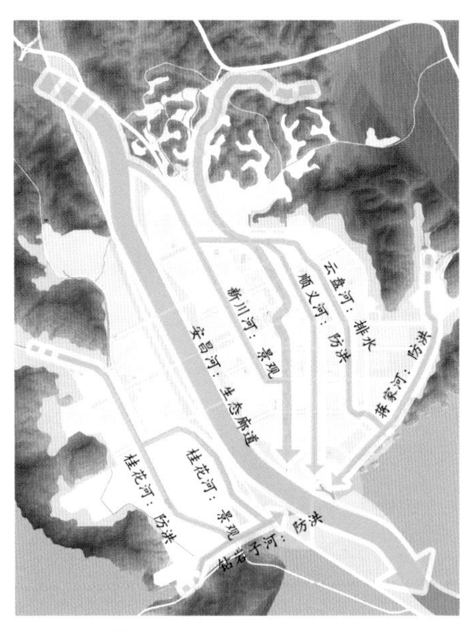

图6 水系格局 Fig. 6 Water network

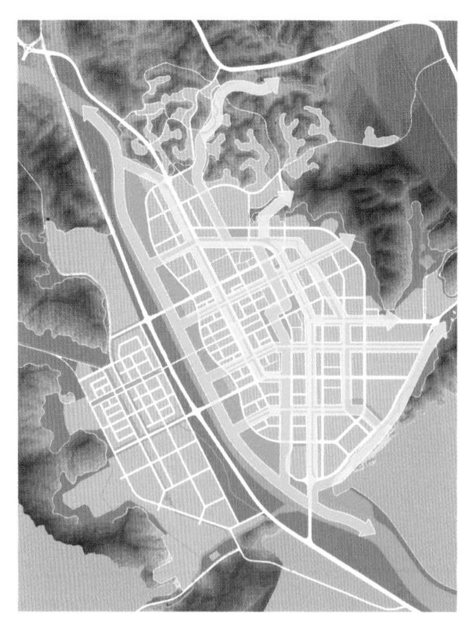
图7 城镇肌理 Fig. 7 Urban pattern

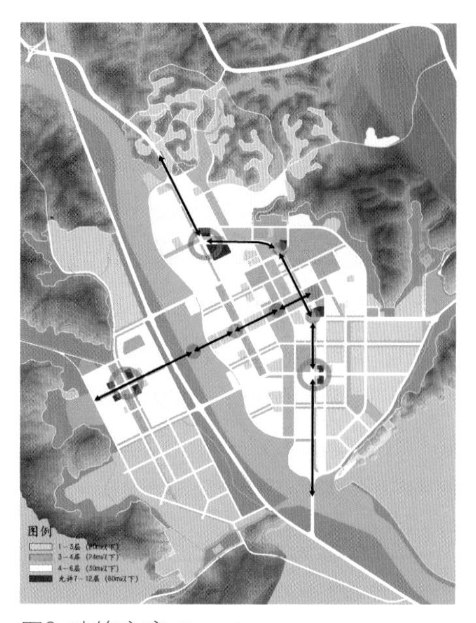

图8 建筑高度 Fig. 8 Building height

造技术、材料、设计手法丰富等种种变化。如何在由“寨”到“城”的转变中实现在文化传承的基础上进行时代性创新是城镇风貌规划设计工作的核心问题与努力方向。

北川新县城城镇空间结构的整体设计着眼于羌族传统聚落的营造传统，着重对城市山水格局、肌理形态、建筑高度提出构想。

1．山水格局——望山融丘，理水亲人

北川新县城城区建设用地主体位于群山环绕的河坝之上，规划对于不同尺度的山采取不同的使用态度：严格保护大型山体，城市建设用地组团拥山而不占山，观山而不登山，保护山体作为城市背景；合理利用浅丘地带，适度进行开发，城市建设用地呈分散组团布局，与丘体呈交错布局，丘融于城，城丘一体（图5）。

水体利用意在传承羌人精于水利、水寨一体的聚居传统，提升新县城城市生活的亲水性：对外以安昌河为主脉打造连接新县城与安昌镇的城市滨水风光带，以景联城；对内在保持原有山体排水和农用灌溉等水系格局的基础上，梳理现有水系，赋予不同区位的水以不同的功能使命，兼顾景观优化与市政安全的双重需求，引水造景，营水富景，形成亲水活动的步行范围不超过300m的城区水系格局（图6）。

2．城市肌理——集约紧凑、疏密有致

结合土地资源、自然环境、建设效率与微观气候等多要素的统筹考虑，基于城市周边山体合理保护利用的原则，有收有放，形成适度紧凑、边界自由的城市平面形态，人均建设用地控制在105m²/人以内（国标GBJ137—90：此类少数民族地区人均建设用地可放宽到150m²/人）；基于城市内部开放空间整体设计的原则，连山通水，形成具有开放空间网络的城市用地组团格局；基于城市内部建设运行效率优先的原则，以双棋盘的路网格局形成相对规整的城市地块划分，有效服务于快速重建；基于城市范围环境气候舒适宜人的原则，城镇开发建设疏密相间，“廊”（安昌河）“管”（垂直安昌河设有绿化带的城市道路）结合，缓解静风效应，促进城市内部空气流动（图7）。

3．空间形态——整体趋缓、局部求变

城镇整体建筑高度借鉴羌族传统营造模式，整体形态保持平缓舒展的态势，契合山水环境保护的要求与城市气质定位。建筑群落以低层、多层错落布局为主，局部点缀小高层。而小高点层建筑布局考虑以城镇干路与城市景观线为主要活动路径，形成点状连续、视觉引导的空间行进体验。在保持片区高度变化平缓的基础上，鼓励以点缀的手法丰富局部建筑轮廓线的变化，形成高低错落的建筑形体感，丰富近人尺度的空间感受（图8）。

三、建筑风貌指引概述

城镇风貌是城镇市气质、底蕴、精神的外在展现，是城镇文化及社会发展程度的综合反映。北川新县城的城镇风貌和建筑风格就是羌族文化延续与发展的外显。城市设计以中国建筑学会等建筑规划学界权威组织为核心，广泛邀请业内知名专家学者就城镇风貌特色以及重点建设项目建筑设计进行多次讨论与评审。2009年7月，中国建筑学会，中国城市规划设计研究院和绵阳市委、市政府联合举办了北川新县城城镇风貌和建筑风格专家研讨会。会议形成六点共识，作为北川新县城建筑设计方案确定的重要原则：

1）作为城建工程标志、抗震精神标志与文化遗产标志，北川新县城城市规划建设更要立意高远，突出民族文化传承与生态环保意识，把打造文化遗产之城与低碳绿色之城作为当前建设和长远发展的目标。

2）城市建筑风貌应以体现羌族建筑文化传统为基础，重视统一性与多样性的平衡，做到统一之中有变化，变化之中保统一。坚持通过规划和城市设计控制来保持建筑风格的整体协调，以成就城市特色，避免过分突出单体建筑而削弱城市整体特色。

城市中的居住建筑等作为城市背景建筑，应体现建筑风格的统一，重要公共建筑则应当重点突出形象与风格的创新。

3）传统是历史川流不息的长河而非一潭死水，建筑设计应在羌族文化传承的基础上强调创新。"形而上者谓之道，形而下者谓之器"，正确处理道与器的关系，道宜一脉相承，而器则与时俱进。

4）建筑的设计应努力实践"乡土建筑现代化，现代建筑本土化"的原则，在满足时代功能需要的前提下力求在建筑的体量、尺度、色彩、材质等方面真实地表达羌族文化，反对简单的抄袭搬用，杜绝表里不一的虚假"化妆"，实现建筑形式与功能的统一。

5）所有建筑应突出绿色设计理念，强调节地、节能与环保。

6）重要文化建筑应争取创造传世之作，力求达成建筑艺术性、美感与震撼力的统一，使人过目不忘；行政建筑应强调稳重、平实、亲民与简约，切忌虚张声势，应经得起社会舆论的检验与挑剔。

在此基础上，北川新县城工程建设指挥部和中规院根据进一步整理出《北川新县城城镇风貌与建筑风格实施指导意见》。这次研讨会成果丰硕，对后期具体方案设计和专家评审起到了重要的指导作用。城市设计对建筑设计的指引与参与在六点共识的基础上，形成了"四求两法"的工作思路与技术策略。

1. 风格原则——四求

1）求特色——北川新县城建筑风格应在应以体现羌族建筑文化传统为基础，通过挖掘北川地区羌族聚落传统的物质文化与非物质文化的民族元素并加以灵活运用，突出建筑的地域特质与文化内涵。

2）求协调——建筑设计对外重视城市环境的营造，与周边建筑、山水环境在建筑高度、第五立面（屋顶）形式、建筑立面划分、材质色彩、细部手法与建筑元素比例等方面寻求呼应，对内应与内部空间与功能需求结合考虑，避免单纯的外观符号化、布景化设计。

3）求丰富——建筑形体高低错落与平坡屋顶穿插组合以及丰富的细部元素是羌族建筑聚落组织与设计的优秀传统手法。在保证大的形体关系协调、材质颜色统一、屋顶形式类似的协调风格的前提下，鼓励通过具体的手法变化强化单体建筑，尤其是转角、对景等重要区位的单体建筑的形象识别性，以丰富近距离、慢速度的城市风貌感知。

4）求创新——建筑设计应当秉承羌族文化融合发展的传统特质，在延续文脉的基础上创新发展，鼓励通过新形式、新结构的设计，塑造地标建筑，通过新材料、新技术的运用，创新建筑表现，通过新理念、新设备的推广，探索绿色设计，通过创新为城市风貌的可持续发展带来活力，弥补大规模建设的时空压缩所带来的缺憾。

2. 技术策略——"两法"

1）风格之法

基于延续羌族建筑文化传统，传承与创新并重的原则，兼顾风貌特色与经济可行的双重考量，规划采取分区、分类控制的策略，以羌式建筑为风格基调，从设计手法上分为原貌体现、精华传承与现代演绎三种，对应风格类型分为原生羌风、传承羌风与现代羌风。规划设计对各风格类型的空间布局、总体要求、建筑形体、包括建筑细部、材质色彩在内建筑要素等均进行了针对性的指引，建设项目的区位特点、功能需要以及重要程度，采取不同的设计手法与风格类型策略（图9、表1、表2）。

2）取舍之法

在城区尺度的建筑风格分区控制的"大统一"的基础上，北川新县城规划设计更重视建筑尺度以及人体尺度的"小丰富"。结合建筑建造经济性的考虑，规划提出了"公建重于居住"、"（街坊）边角重于内部"、"（建筑）底、顶重于中段"、"小品重于建筑"的四个"重于"的微观层面的风貌设计策略（图10），基于人对于建筑环境感知与接触部位强弱

表1 建筑风格手法对应表

设计手法	风貌类型	设计定位
原貌体现	原生羌风	焦点建筑，主题街区
精华传承	传承羌风	背景建筑，建设主体
现代演绎	现代羌风	焦点建筑，点睛之笔

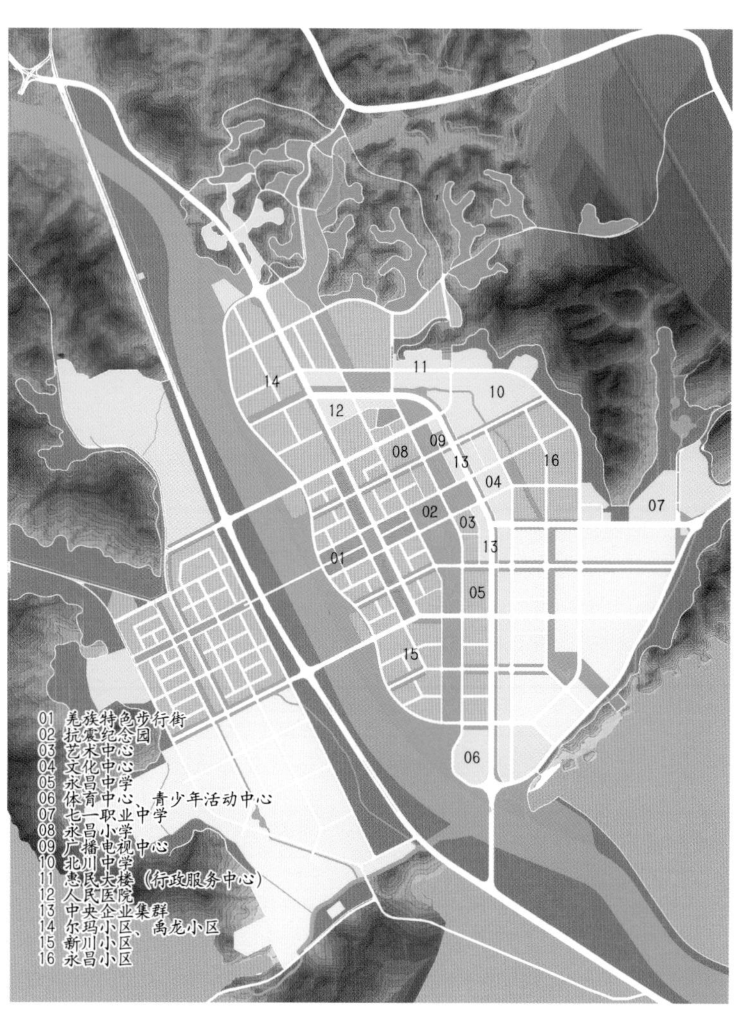

原生羌风型 设计单位：清华城市规划设计研究院

传承羌风型 设计单位：中国建筑设计研究院

现代演绎型 设计单位：中国建筑设计研究院

图9 建筑风格分区分类控制 Fig. 9 Architecture styles layout

公建重于居住
Prioritize public buildings over residential

边角重于内部
Prioritize exterior of building blocks over interior

底顶重于中段
Prioritize the top and bottom of a building over the middle

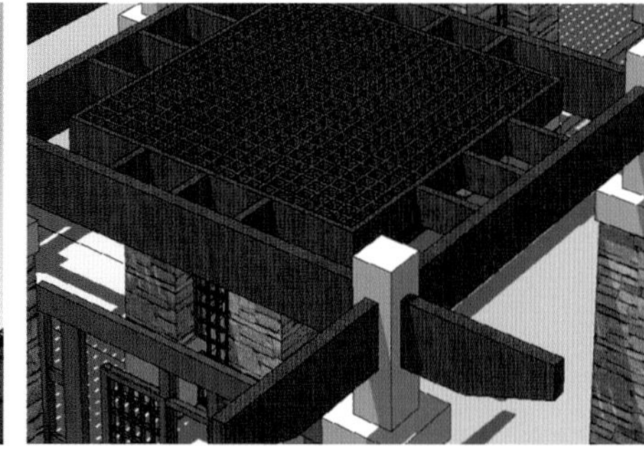

小品重于建筑
Prioritize street furnitures over buildings

图10 取舍之法 Fig. 10 Design strategies

表2 建筑风格手法指引表

建筑风格指引	原生羌风型	传承羌风型	现代演绎型
空间布局	位于城市核心区域的特色商业街以及北部低丘片区	城市建筑的主体类型，以住宅建筑为主，涵盖除重点公共建筑以及工业仓储建筑外的所有建筑	主要包括行政、文化、教育类的城市级重要公共建筑，分散于城市各重要空间节点，是城市建筑文化创新的重要载体
总体要求	满足设计功能需求的前提下，严格按照传统手法设计与建筑选材建造传统羌式建筑聚落	融合现代建筑科学技术与传统羌族建筑文化，在形式与功能统一的基础上，形成整体化的建筑风貌片区	鼓励从建筑形式、材料选择到空间手法、建筑技术的全方面创新设计，强调“现代建筑的本土化”
建筑形体	以3 层以下的低层建筑为主，建筑高度不超过16m（特殊构筑物，如碉楼除外）；强调建筑形体局部与屋顶面的体块高低错落、凸凹变化，形成丰富的建筑立面与屋脊线；鼓励以过街楼的形式实现单体建筑间的沟通与整合	以4～6层的多层建筑为主体，建筑高度不超过24m；建筑形式建议以三段式为主(底层、中段与屋顶)；平面布局强调一定的错落围合关系。鼓励屋顶设计平坡结合	建筑高度可结合建筑设计而灵活确定，应避免过于突兀、单一的巨大建筑体量出现；大型建筑建议以建筑群体的形式出现，保持自身形象协调和谐的同时更应重视与周边建筑在体量上的协调与过渡
建筑要素	建筑形式、材料、色彩、细部力求准确反映羌族传统建筑原貌，材质以石、砖、木为主，颜色以石材的灰色调为主，点缀以木材的暖色与适度的白、黄。注重细部刻画，以白石、檐线、晒晾木构架、花式窗框等传统形式丰富建筑立面，强化风貌特色感知	建筑形式、材料、色彩、细部等要素应在街坊范围内保持整体统一。建筑色彩建议基座采用浅灰、浅褐色，主体采用米白、鹅黄色，屋顶采用青灰色，局部点缀以木色等暖色。建筑底层、屋檐下部、门窗外框是细部刻画的重点，建议进行适度的羌式装饰处理和民族元素点缀	鼓励新材料、新技术的运用，但建筑的色彩材质等视觉观感应与传统材料相接近，和城市环境相协调，门窗等建筑元素建议以传统对应元素为原型，进行适度的抽象、提炼，避免不顾尺度的变化而简单放大、生搬硬套传统民族元素
其他要求	建筑设计应考虑广告店招的摆放，鼓励特色店招；严禁空调外挂机等建筑附属物的视觉干扰	设计应考虑底层商业店招设置与建筑附属物的隐蔽设计；沿街建筑禁止使用外露安全网与金属卷帘门；街角建筑需强化建筑转角的视觉特征	建筑设计应整合景观设计，统筹考虑室内外空间的过渡，建筑元素与景观元素的统一与呼应

的考虑，在城市、街坊、院落以及单体建筑等不同尺度层面，强调风格刻画的有取有舍、不均用力，塑造城市风貌不同尺度的丰富性。

3．工作方法

单体建筑设计是总体规划和总体城市设计的延续，是塑造城镇风貌的重要载体与表现终端。北川新县城单体建筑设计针对不同项目的特点和时间安排，采取直接委托、方案征集相结合的灵活方式。对于量大面广、政策性和技术性较强的安置住房、市政工程和园林景观设计项目采取直接委托的方式；对于重要的公共设施项目，如学校、医院和文化设施等，主要采取方案征集的方式，并通过相应的制度设计将设计协商与实施管理融为一体。

1）住宅建筑设计委托

为了切实贯彻民生优先，率先开展保障性住房建设，北川县在2009年2月委托中国建筑设计研究院和中国城市规划设计院（部分建筑方案设计委托青岛市建筑设计院合作），承担征地拆迁安置房和受灾群众安居房修建性详细规划和建筑方案设计工作，并专门制定了北川新县城住宅设计导则和统一技术措施，较好的引导了安置住房施工图设计和工程实施。

2）公共建筑方案征集

为指导公共建筑设计方案征集工作，北川县政府于2009年5月正式出台了《北川新县城政府投资工程建筑设计方案征集暂行办法》，成立了新县城建筑设计方案征集委员会，由北川县政府、中国城市规划设计研究院、山东援建北川工作指挥部和业主单位共同组成，负责方案征集工作。

方案征集主要采取邀请征集方式，每个项目邀请3至6家建筑设计单位，包括援建方山东省、四川省内优秀建筑设计单位，以及其他国内外优秀建筑设计单位等50余家甲级资质建筑设计单位。

方案评审邀请包括科学院、工程院院士、业内设计大师在内的国内一流的建筑设计专家进行方案评审。每个项目邀请5～7名专家，方案征集委员会各方代表参加会议，纪检监察和审计部门到场监督，确保方案评审公平、公开和公正。评审专家从征集方案中推荐1至2个推荐方案，经过方案公示后，提交城乡规划委员会讨论决策。

3）城市设计实施控制

为了科学、系统、高效地统筹推进新县城建设，北川新县城建立起北川新县城工程建设指挥部、山东援建北川工作前线指挥部、中规院北川新县城规划工作前线指挥部，完善了"三指联系协调工作机制"，规划设计坚持中规院技术总把关，中规院专家组全面负责拟定规划条件和城市设计导则，参与专家评审，负责规委会审定方案的修改完善，参与审查初步设计和施工图设计，确保设计质量。

为了保证建筑工程施工符合城市设计和建筑方案设计要求，北川新县城建立起施工图规划复核、建筑外立面样板墙审查规划巡查制度等制度。从方案设计到初步设计，再到施工图设计，做到规划全程把关，确保城市设计意图最终贯彻落实。在外立面装修施工以前，要求设计单位必须在施工图基础上进一步细化深化外立面装修的详细要求，包括色卡、材料、详细做法，作为指导施工单位实施外立面的重要依据。在整个过程中充分体现了规划师与建筑师的互动与合作。

图11 新县城鸟瞰 Fig. 11 Bird view of the new county seat of Beichuan

Building a New Beichuan
——Planning Design and Townscape for the New County Seat of Beichuan

Zhu Ziyu / Project Supervisor of Comprehensive Planning for the New County Seat of Beichuan
Vice-Commander of the Front-Line Headquarters of Planning for the New County Seat of Beichuan, CAUPD
Li Ming / Project Supervisor of Urban Design Master Plan for the New County Seat of Beichuan, CAUPD
He Wang / Deputy Commander of the Construction Headquarters for Beichuan New County, Deputy Director of Mianyang Urban Planning Bureau
Sun Tong / Project Supervisor of Regulatory Detailed Planning for the New County Seat of Beichuan
Commander Assistant of the Front-Line Headquarters of Planning for the New County Seat of Beichuan, CAUPD

Beichuan, located west of Mianyang city in Sichuan Province, is the only Qiang-ethnic autonomous county in China. Qushan Town, Beichuan's county seat, is situated by the Qian river, at the heart of Longmen mountain peaks. Being a small town, Qushan was not well known by the public in the past and its economy is not quite developed; but it has beautiful scenery, peaceful life, a harmonious social order and a strong Qiang-ethnic culture. The small town, however, due to its location at the intersection of Yingxiu and Leigu, the two earthquake fault zones of Beichuan, has a huge hidden risk. At 2:28PM on May 12, 2008, an 8.0-magnitude earthquake hit southwest China. The disaster turned the once beautiful Qushan Town into a rubble—landslide strike, buildings collapsed, bridges crumpled, and roads caved in. Qushan Town was the most heavily hit area in the 2008 Wenchuan Earthquake (Fig. 1).

Qushan Town sat on grounds with high risk of geological disaster. Confined by its surrounding mountains, the original site of Qushan Town is unsuitable and impracticable for reconstruction. Local residents affected by the disaster also expressed strong wishes to move to a safer, more outstretched and convenient area. On May 19, 2008, under the request of Mianyang Municipal Party Committee and the Municipal Government, the State Ministry of Housing and Urban-Rural Development dispatched an expert team to Beichuan to select a new location for building a new Beichuan. Commissioned by the Ministry, CAUPD led the Post-quake Reconstruction Site Selection and Planning Research for New Beichuan, which put forward the feasibility for relocation of the county seat and proposed alternative sites based on scrutinizing their geological and spatial conditions, taking into account future regional development and possible adjustment of administrative boundaries, in a joint effort with Wuhan Geotechnical Engineering and Surveying Institute, Mianyang Urban Planning Bureau, the CPC of Beichuan Autonomous County and the People's Government of Beichuan Autonomous County. Approved by the Central Government, the new county seat is now located southeast to the town of Anchang, in a valley flatland of approximately $10km^2$ (Fig. 2).

The new county seat of Beichuan is the only relocated county approved by the central government after the May 12th Wenchuan Earthquake, and a unique and new autonomous county for the Qiang nationality in China. Its planning and construction offers a valuable historical chance of cultivating and innovating the Qiang nationality culture, and it shoulders an important mission of showcasing achievements of post-quake reconstruction and the spirit of earthquake relief, and of exhibiting the ethnic and geographical features of the Qiang nationality. As early as May 14, 2008, two days after the 5.12 Wenchuan Earthquake, CAUPD went to the disaster site for relief work. After finishing the prophase relief settlement, CAUPD began to make site selection, comprehensive planning, urban design master plan and infrastructure master plan for the new county seat of Beichuan. In the meantime, by collaborating with more than 50 planning and design organizations, CAUPD has finished a series of planning and design works including town road engineering design, public construction schematic design, schematic design for affordable housing projects, and green space landscape design. By using technical means, it intends to interpret the principles of "safety, livability, characteristic, prosperity, advancement, harmony" and the three symbols, i.e. making the county seat "a symbol for urban construction and engineering, a symbol of earthquake relief, and a symbol of culture heritage", among which the planning and design of townscape is one of the most significant links of the whole series.

I. Overview of Comprehensive Planning for the New County Seat of Beichuan

The comprehensive planning for the new county seat of Beichuan is based on the central government's calling of "building a new Beichuan", the principles of "safety, livability, characteristic, prosperity, advancement, harmony", the three symbols by the State Council, and the principle of "government guidance, expert instruction, department cooperation, public involvement and scientific decision". The planning's core missions include resettling disaster-affected people, restoring town functions, and making a spatial framework for long term development of the county. The planning and design apply "Six Synergies", namely, innovative ideas and practicable approaches, planning strategy and project construction, sector planning and engineering construction, planning regulation and construction management, government decision-making and residents expectation, and plan layout and urban design. The reconstruction should be implemented smoothly in collaboration of multiple specialties.

The comprehensive planning for the new county seat of Beichuan defines that the future new county will be integrated with the development layout of Mianyang, Jiangyou, An County and Beichuan. By 2020, the new county seat will be a political, economic and cultural center of Beichuan, a tourism base of west Sichuan Province, an industrial base of west Mianyang, and a modern Qiang ethnic culture town and an eco-town. In the near future, it will have a land area of $5km^2$ and a population of 35,000; in the long future, it will have a land area of 7 km^2 and a population of 70,000. Based on the urban development objective and the existing natural conditions, the future new county seat will be composed of four functional zones, i.e. Shandong Industrial Park in the south, the integrative living area in the middle, the recreational tourist vacation area in the north and future expansion area in the west. Shandong Industrial Park in the south focuses on industrial development and settling employment for residents. It will include different industrial types which conform to the local resources and condition, protect natural environment, and extend the former industrial base. The park can provide as many as 10,000 job opportunities, and by setting complete supporting services and vocational schools, the local residents can quickly acquire employment skills and posts. The integrative living area in the middle is flat with sufficient water, which is suitable for urban life and public services. By 2010, a total number of 9,000 affordable housing units will be built for disaster-affected people, part of land-deprived peasants and local relocating villagers of Qushan Town; by 2015, additional 2,600 affordable housing units will be built, and the building area of affordable housing units will account for 46% of the total housing building area by the end of the planning. Due to complex landform, disperse land for use, sensitive eco system and beautiful environment as well, the recreational tourist vacation area in the north will mainly focus on recreational and cultural industries as well as welfare and rehabilitation services (Fig. 3).

The comprehensive planning for the new county seat of Beichuan proposes a spatial structural design idea of "landscape circle, ecological corridor, leisure belt, growth ridge, service chain and landscape axis", and these multiple spatial elements intertwine and form a spatial frame of spatial experience and townscape exhibition, which can provide definite, orderly and efficient layout for urban reconstruction. In such a layout surrounded by mountains and rivers, the Anchang River serves as an "eco-corridor" from south to north of the new county, and the greening area on both banks of the river can improve environment and ensure microclimate quality in the area with high frequency of static wind, which becomes a major eco-corridor of the new county and also a main site for leisure and fitness as well. Most public buildings of the county stand along the road skeleton and form a public service chain where the Cultural Center, Earthquake

Memorial Park, Livelihood Building, schools, hospitals and Youth Center are located. The "leisure belt" along the Yongchang River is composed of the former water system and a series of special parks, which integrates historical relics, outdoor activities for the youth and elderly, earthquake memorial, sports and recreation with the riverside park in an organic way. The "landscape axis" which skillfully utilizes the surrounding mountain shapes and river trend in a west-east direction can unify the whole urban functions and spatial landscape. Lying against Yuanbao Mountain, Tazi Mountain and Yunpan Mountain, the landscape axis connects not only the mountain and river space, but also the Cultural Center, Memorial Park, the Qiang Features Pedestrian Street and the pedestrian bridge, forming a key walkway of the county seat and becoming a backbone for spatial structure and functional structure of the new county seat of Beichuan (Fig. 4).

II. Overview of Urban Design Master Plan for the New County Seat of Beichuan

As the new county seat of Beichuan is a fresh new town and the only autonomous county of the Qiang nationality, the inheritance and rehabilitation of the local ethnic culture are undoubtedly the origin of design. Over the past hundreds or even thousands of years, with the blending of the native Qiang culture and alien cultures, the traditional Qiang settlement has gradually formed its own organic and natural space creation essence which is in line with its local conditions and makes best use of water. Its spatial layout and form are not restricted by conventional way and method, and there formed an integrated, unified, environmentally harmonious settlement townscape by virtue of imitating construction using the same material and process. What challenge building and creation of distinctive county seat of Beichuan are various kinds of changes, including change from tradition to reality, change of geological environment from village to county, expansion of scale, complex function organization, change of life style from an introvert and defensive type to an extroverted one with frequent social interaction, enhanced building technique, materials and design skill, etc. Therefore, the core direction of townscape planning and design lie in how innovation can be realized based on tradition heritance in the process of transforming from a "village" into a "town".

The overall spatial structural design for the new county seat of Beichuan has its focus on the traditional Qiang settlement, concentrating on conceiving landscape layout, texture form and building height.

1. Natural scenery pattern: an intimacy with nature

The main construction land for building new county seat of Beichuan lies on a river flat land encompassed by mountains. The natural landscape of the surrounding mountains is preserved by concentrating development in the valley area with an unobstructed view of the beautiful and encircling mountain range. The preserved mountains can serve as suburban experience park to enrich public activities; the foothill areas are moderately developed to create a city-mountain integrated cityscape (Fig. 5).

By inheriting the Qiang-ethnic tradition of sophisticated water use, easy access to waterfronts is designed for all residential areas: out of the town, to create a waterfront scenic belt linking the new county seat of Beichuan and Anchang Town with the Anchang River as its main context, connecting the counties with landscape; inside the town, to improve the existing water system by retaining the existing drainage and irrigation patterns of the mountains and water system, and endow different areas with different functions and missions; by taking account of the dual demands of landscape optimization and municipal safety, to create landscape by introducing water and enrich the landscape by making the best use of water, thus forming a water system layout of "four rivers" (the east bank area) which people have easy access to water by walking within 300m (Fig. 6).

2. Structural pattern: a compact arrangement with proper density

By taking into account of factors including land resources, natural environment, building efficiency and micro-climate, the townscape planning and design strive to create an integrative cluster layout which is properly compact and free in boundary with regular pattern and density. Firstly, according to the principle of rationally protecting and utilizing surrounding mountains of the town, a compact urban plane form with free boundary is formed with a per capita construction land under 105m^2/person (the national standard GBJ137-90: the construction land per capita for ethnic areas can be 150m^2/person); secondly, mountains and water should be linked by following the principle of making a general design of the inner space of the town to form a grouped layout based on the open space network of the town; thirdly, based on the principle of giving priority to construction and operation efficiency inside the town, the town should have unified texture, and its plots should be divided in a relatively regular way through road network pattern as a dual chess board, which can serve for quick reconstruction; finally, based on the principle of creating a comfortable and pleasant environment and climate in the town, the town development and building should have proper density which combine "corridor" (the Anchang River) and "gallery" (the town road with a green belt vertical to the Anchang River), so as to ease static wind action and facilitate internal air flow and dispersion of pollutants in the county (Fig. 7).

3. Spatial form: an integral with local variation

The overall urban form maintaining a smooth and outstretched pattern, in accordance with the small-town quality and the requirement of preserving the natural environment; and low-rise and medium-rise buildings mostly mixed together and a few high-rise buildings located along major roads and landscape corridors. While keeping a gentle height variation in the whole area, it is allowed and encouraged to enrich the local building contour line with different height of buildings, so as to form diversified building shapes, pass on traditional building methods, and enrich the human scale (Fig. 8).

III. Overview of Townscape Design

Townscape is external presentation of the spirit, style and inherent quality of a town, and also an overall reflection of the cultural and social development level of a town. The townscape and building style of the new county seat of Beichuan function as external presentation of continuity and development of the Qiang culture. The townscape design organizes authoritative institutions including Architectural Society of China to invite a wide range of renowned experts and scholars to make multiple discussions and reviews on the townscape and planning and design for key projects. In July 2009, an expert seminar on townscape and building style of new Beichuan was jointly held by Architectural Society of China, China Academy of Urban Planning and Design, Mianyang Municipal Party Committee and Mianyang Municipal Government, on which six point of consensus were reached as major principles for the architectural design scheme of the new county seat of Beichuan:

1) As a symbol for urban construction and engineering, a symbol of ,earthquake relief and a symbol of culture heritage, the town planning and construction of new Beichuan should stand on a high level to stress the ethnic and cultural inherence and environmental protection, and make its current and long term development goals as building a cultural heritage town and a low-carbon green town.

2) The townscape should base on reflecting architectural and

Table 1 Architectural Style and Corresponding Design Technique

Design Technique	Style	Positioning
Recreation	Origina	Focal buildings, theme blocks
Adaptation	Adapted	Background buildings & main construction
Modernization	Modern	Focal buildings, landmark

Table 2 Table of Architectural Style and Method

Architectural style	Original	Adapted	Modern
Spatial layout	Commercial strip in the core area of the town and low hilly area in the north	The majority of buildings in the town is residential buildings, which covers all buildings except key public buildings and industrial warehouses	It mainly includes important administrative, cultural and educational and other public buildings, which are discrete in various important spatial nodes and are important carriers of architectural culture innovation of the town
General requirements	On the premise of meeting the designed functions, to build the traditional Qiang settlement by strictly following the traditional skill and building material selection	To incorporate modern architecture science & technology and the traditional Qiang architectural culture, and to form an area with integrative architectural style based on unified style and function	To encourage all-round innovative design from building form, material selection to spatial technique and building technology and stress "localization of modern buildings"
Building shape	The majority is three-storey buildings and below, and the building height should not exceed 16m (excluding special structures such as turret); to highlight discrete local building shape and convex & concave roof top, form rich building elevation and ridge line; communication and integration of single buildings through arcade is encouraged	The majority is four-to-six-storey buildings, and the building height should not exceed 24m; the suggested building form is mainly a three-section one (the ground floor, middle section and roof); in plan layout, discrete and enclosed relationship is stressed and a combination of flat and pitched roof is encouraged	The building height should be determined by considering the architectural design, and single huge building mass should be avoided; it is suggested that large building be designed in the form of building complex, maintaining its own harmonious image while stressing harmony and transition to surrounding building in terms of mass
Architectural elements	The building form, material, color and details strive to accurately reflect the original appearance of the traditional Qiang building, mainly adopting stone, brick and timber, with gray stone as major color, added with warm color of wood, white and yellow. Details are given attention to, building elevation is enriched by traditional styles such as white stone, eave line, sunned timber frame and patterned window frame, and a perception of townscape characteristics is strengthened	The blocks should have the same building form, material, color and details. It is suggested that building bases adopt light gray and light brown; main body adopts beige and light yellow; caesious roof with local warm colors such as timber color. The key of details are building ground floor, part below eave and external frame of door and window, and it is suggested to have appropriate Qiang style decoration and national elements	The use of new materials and new technologies is encouraged, but visual perceptions such as the color and materials of building should be close to traditional materials, and be in harmony with urban environment. Based on traditional elements of door and window, building elements should be abstracted and refined to avoid simply enlarge and copy traditional elements regardless of changes in scale
Other requirements	Arrangement of shop logos should be considered in the architectural design, special shop logos with are encouraged; it is prohibited to have visual disturbance by some building auxiliaries such as air conditioning outdoor units	Arrangement of shop logos on the ground floor and concealed design of building auxiliaries should be taken into account; it is not allowed to use exposed safety net and metal shutter door for building facing streets; buildings at street corners should have strong visual feature	The architectural design should integrate landscape design, transition between the internal and external space should be taken into full consideration, and architectural elements should be in harmony with landscape elements

cultural traditions of the Qiang nationality, emphasize balance of integrity and diversity, and make the town have different faces in the integrity. Planning and town design should be utilized to control the integral coordination of building style to showcase the town features and avoid weakening the overall features because of extremely highlighting single buildings.

Residential buildings, as background of the town, should have a uniform building style, while major public buildings should reflect innovation in terms of image and style.

3) As tradition is an unceasing river of the history, the building design should emphasize on innovation on the basis of the Qiang cultural inheritance. Design theory is metaphysical while design outcome is concrete and specific. It is important to deal with the relation of the two to make theory be uniform with the origin and make outcome be upgrading with the times.

4) Building design should implement the principle of "modernization of local native buildings and localization of modern buildings". After satisfying the functional demands of the time, a real Qiang culture should be presented in volume, dimension, color, material and other aspects. Building form and function should be uniform while simple imitation, copying, or untrue "dressing up" for

buildings is prohibited.

5) All buildings should highlight the green design concept focusing on land saving, energy saving and environmental protection.

6) For major cultural buildings, master pieces are advocated to last for centuries which are unforgettable and harmoniously integrating artistic, aesthetic and inspiring power. For administrative buildings, they should be solemn, amiable and concise which are not arrogant and can stand examination and criticism of the picky opinions from the society.

Based on these aspects, the Construction Headquarters for the New County Seat of Beichuan and China Academy of Urban Planning and Design further compiled Guiding Opinions on Townscape and Building Style Implementation for the New County Seat of Beichuan. As fruitful as the seminar is, it plays an instructive role for design of the detailed scheme and expert appraisal. The guidance and involvement of urban design on building design, based on the six points of consensus abovementioned, form methodology and technical strategy in "four pursuits and two methods".

1. Four principles of architectural design:

1) Distinctiveness: when reflecting the architectural culture tradition of the Qiang nationality, the architectural style of the new county seat of Beichuan also highlights the geological characteristics and cultural connotation by exploring traditional tangible and intangible cultural elements of the Qiang settlement in Beichuan and applying them flexibly [3].

2) Harmony: externally, the architectural design focuses on creation of urban environment and seeks harmony with the surrounding buildings and landscape in terms of building height, the fifth façade (roof) style, building façade division, materials and colors, details, and architectural elements, etc.; internally, it takes account of the internal spatial and functional demands to avoid simple symbolization and settings in designing the appearance.

3) Variety: the excellent traditional skill of the Qiang settlement in building organization and design include discrete building shape and pitched roof as well as abundant details. While ensuring that the general shape relationship is harmonious, material and color is unified and roof has similar type, it is encouraged to make the image of single buildings, especially those in important positions such as corners and opposite sceneries identifiable by virtue of detailed skill variation, so as to consolidate perception of townscape in a short distance and at a low speed.

4) Creativity: the architectural design should carry on the traditional characteristics of the Qiang culture, and innovate while continue the cultural elements. It is encouraged to create landmark buildings through design of new styles and new structures and by using new materials and technologies.

2. Design Technique

1) Style

Based on the Qiang architectural culture tradition, guided by the principle of laying equal stress on inheritance and innovation, and by taking into account townscape and economic feasibility, the planning adopts the strategy of zoning and classification control. With the Qiang style building as the basic style, three design techniques, namely, recreation, adaption and modernization, are used, and the corresponding styles are the original, adapted and modern. The planning and design have corresponding requirements on spatial layout, general requirement, building shape, architectural elements including details, materials and colors. Different design skills and style strategies are adopted based on different location features, functional requirements and level of importance (Fig. 9, Table 1, Table 2).

2) Way of adoption or rejection

While the architectural style is controlled by zonings and "unified in general", the planning and design of the new county seat of Beichuan attach more importance to enrich building scale and human scale. Enrich architectural diversity through the following strategies: prioritize public buildings over residential, prioritize exterior of building blocks over interior, prioritize the top and bottom of a building over the middle, and prioritize street furnitures over buildings (Fig. 10). Considering people's sense of building environment and contact parts, style adoption or rejection should be stressed in aspects of city, block, courtyard and single building, so as to mold richness of townscape with different scales.

3. Methodology

Single building design is an extension of comprehensive planning and urban design, which is a main carrier of cultivating townscape and a terminal of presentation. The single building design for the new county seat of Beichuan applies flexible means such as a combination of direct consignment and scheme recruitment for characteristics and time schedule of different projects. For resettlements which are large in quantity and area with strong political and technical features and municipal engineering works as well as park landscape works, direct consignment is adopted. While for major public services projects such as schools, hospitals and cultural facilities, the method of scheme recruitment is applied, and by using corresponding system design to integrate design negotiation with construction management.

1) Design consignment for residential building design

In order to fully reflect the priority of citizens and take the lead to conduct affordable housing construction, in February 2009, Beichuan County consigned CAG and CAUPD (part of architectural scheme design was consigned to Qingdao Institute of Architectural Design for collaboration) to make detailed plan and architectural scheme design for resettlements of land acquisition for removal and for resettlements of disaster-affected people, and to formulate guiding principles and uniform technical measures for designing residence in the new county seat of Beichuan, which can play an instructive role in making construction drawings design and project construction for resettlement housing.

2) Design scheme recruitment for public buildings

In May 2009, the county government of Beichuan officially issued *Interim Methods of Recruiting Design Schemes for Government-funded Projects in the New County Seat of Beichuan* to instruct the design scheme recruitment for public buildings. On that occasion, a committee of design scheme recruitment was established which is composed of the government of Beichuan Qiang Autonomous County, CAUPD, Headquarters of Projects aided by Shandong Province and the owners to take charge of all related work.

An inviting-type recruitment was basically applied for design scheme: 3 to 6 designing institutions were invited for each project among more than 50 architectural designing institutions with Grade A qualification from the aiding province Shandong, excellent designing units in Sichuan Province and other renowned designing companies from home and abroad.

At the same time, first-class architectural design experts including academicians of Chinese Academy of Science, Chinese Academy of Engineering and design masters in China were invited for scheme appraisal. Each project invited 5 to 7 experts and representatives from the committee of design scheme recruitment participated in the sessions with site supervision by the discipline inspection and supervision authority and auditing authority to ensure fairness, equality and openness of the scheme appraisal. The appraisers recommended 1 to 2 schemes among the candidates, and after public notification, the selected schemes were submitted to the county planning committee for discussion and final decision.

3) Implementation control of town design

In order to proceed with construction of the new county seat in a scientific, smooth and highly efficient way, a "three-headquarter coordination system" was built which included the Construction Headquarters for the New County Seat of Beichuan, the Front-Line Headquarters of Aiding Projects to Beichuan by Shandong Province and the Front-Line Headquarters for the New County Seat of Beichuan Planning by CAUPD. The planning design insisted on the working principles that CAUPD was in charge of general technical control in which the experts team from CAUPD prepared planning conditions and guidance of town design, involved in the appraisal, modified and improved the scheme approved by the planning committee and involved in reviewing the preliminary design and construction drawings design to ensure design quality.

Besides, several systems such as recheck system of construction drawings and inspection system of model walls for building façades were also set up to ensure that project construction can conform to the requirements of townscape design and architectural design scheme. The whole process, from scheme design, preliminary design to construction drawings design, was strictly controlled so that the intent of townscape design can be finalized. Prior to the finish construction of façades, it was required that the designing company make detailed conditions of the façade finish works on the basis of the construction drawings including color pantone, materials and detailed schedule to make them as significant basis for contractors in building the façades. Planners and architects were interacting and coordinating well throughout the process.

从云盘山眺望新县城（林永新摄）

Overlooking the new county seat from the Yunpan Mountain (by Lin Yongxin)

从狮子山眺望新县城

Overlooking the new county seat from the Shizi Mountain

禹王桥
Yuwang Bridge

安居北川

刘燕辉 / 中国建筑设计研究院副总建筑师

住房建设是灾后重建的关键所在，直接关系到人民群众的切身利益，最能体现对灾区人民的关爱，也是考量抗震救灾成就的重要指标。因此，新北川的安居工程被定位为整个新县城建设的“重中之重”。

新北川居住建筑包括原北川县曲山镇受灾群众的安置住房和新县城用地范围内失地农民的拆迁安置住房两部分，共计100多万平方米。

在居住建筑设计中着重体现了以下六项原则：

一、贯彻十二字方针的原则

“安全、宜居、特色、繁荣、文明、和谐”十二个字方针是对新北川建设的总体要求，对居住建筑而言，具有直接和至关重要的作用，它是指导安居房设计的基本依据。

根据北川居民的生活习惯，安居房住区的结构组织以小街坊为基本单元，空间开放，步行系统连贯；街坊的尺度较小，有助于增加归属感，形成亲密的邻里关系。同时，街坊结合自然景观，使住宅能够享有较多景观资源。

北川新县城更是一个具有“特色”的新城，在安居房建筑设计上充分考虑到了对羌族文化的延续和传承；在建筑风貌上主要吸收羌族“白石崇拜”的传统，提取汉羌建筑符号，通过传统建筑材料和现代建筑结构的有机结合，塑造出羌风浓郁的当代建筑风貌。

二、贯彻社会公平原则

安居房的居住对象为地震受灾群众以及拆迁农民，由于涉及切身利益，他们对住宅的关心程度非常高，只有体现社会公平原则，才能更好地落实党和国家抗震救灾的一系列政策。

为了使群众满意，政府放心，我们根据项目特点制定了具有针对性的技术路线，在规划前期进行了深入的调研工作，这对保证社会公平起到了重要的作用，同时指导了居住建筑设计。在居住建筑设计中严格按照政策的指导方向，对具体的户型面积以及配置进行了严格的设计核对，针对不同类型家庭的人口构成特点，在温泉片区一期共设计了50m^2、70m^2、90m^2、105m^2、120m^{2}多种户型单元，误差精确到1m^2。并按实际需要配置了储藏间，室外生活阳台，明厨明卫等必要功能和条件。

根据北川新县城安居房安置办法，为使居民尽快得到妥善安置，还对安居房进行了包括室内瓷砖地面、室内门、厨卫墙砖、经济型节水洁具、厨房灶台、内墙涂料等在内的装修设计，使居民不需二次装修就可直接入住，在细节上体现了宜居建筑的设计理念。住宅底层尽量设置了小型商铺，充分地满足了居住区内的就业及经营需要，也为居民提供了充足的服务设施。

三、贯彻“羌风羌貌”的原则

北川是我国唯一的羌族自治县，重筑羌家新城得到各界的共识。“羌风羌貌”不仅可以塑造新北川的形象，也是创造旅游资源的重要元素。根据北川新县城总体规划的要求，新县城属于汉羌风貌区，住宅形象设计充分尊重地域文化特色，借鉴民居建筑特点，结合气候特征与场地条件，体现了传统建筑特征，反映地域文化。

住宅采用平坡相结合的屋顶形式，底层为外装饰面砖，顶层为木色涂料，从材质和颜色上体现木石结合的地域建筑特征。同时在坡屋顶和平屋顶的组合上，采取周边和院落内的区分方式，丰富空间组织，降低造价，加快工期。

四、贯彻“红花绿叶”原则

面对一个新城，同时同地大量崛起的建筑，为更充分表现羌风羌貌，确定了“红花绿叶”原则。即：对于整个北川新县城而言，公共建筑应该是“红花”，居住建筑定为绿叶；对于每个居住小区而言，沿河、沿街的建筑是“红花”，组团中心的建筑定为“绿叶”；对于

安居房在北川新县城中的位置

Location of affordable housing quarters in the new county seat of Beichuan

住宅而言，环境和小品是“红花”，住宅建筑定为“绿叶”。正是由于“红花绿叶”的合理搭配，使整个城市更加层次分明，突出了重点标志性建筑，也营造了闹中取静的居住氛围，同时也在节约投资，加快工期等方面起到了积极作用。

住区内景观设计也是坚持了“红花绿叶”的设计原则，注意采用成本较低的材料，“粗粮细作”：如采用不同颜色的透水砖铺砌出羌绣图案，收到了良好的效果。从羌族人民的生活入手，以羌族物质与非物质文化、民风与民俗为主线，在结合地域文化的基础上，倡导与羌族人民密切相关的日常生活及精神生活相融合。在设计中，本着传承与弘扬民族文化的精神，将羌族一年12个月中的主要节日庆典融入景观设计中，从而在景观层面反映了地域文化的延续，丰富了住宅区的人文景观，为羌族居民喜闻乐见。

五、贯彻节能、省地、环保的原则

节能、省地、环保是我国大力提倡的国策，在北川新县城安居房建设中得以体现。

在规划设计中，通过合理的规划布局和道路组织，在保证居住舒适，活动功能齐全的同时，提高了土地的

使用效率。组团式布局增加了道路密度，方便了居民出行，小区绿地景观充分利用周边新县城的景观带，提高居住的绿化质量。安居房的布局满足采光、通风、日照的要求，严格执行了国家和地方标准规范，体现了以政府作为主导建房的严肃性。

在住宅建筑设计上遵循《四川省居住建筑节能设计标准》、《绿色建筑评价标准》等文件，力图创造舒适宜人的居住氛围。设计前期编制的《北川新县城居住建筑设计导则》和《北川新县城居住建筑设计统一技术措施》均以绿色建筑为基准，从选材到住宅功能布局，从构造节点到施工控制，均以建设绿色建筑为目标，使整个建设过程保持在受控状态下。

北川安居房为"交钥匙"工程，新建住宅满足基本入住要求，居民收房以后即可开始正常生活，装修标准也在设计时做了统一规定，这样减少了毛坯房带来的装修浪费，减少了材料垃圾对环境的不良影响，有效满足了各级政府按期完工的要求。

六、建立国家救灾机制的原则

地震是突发性灾害，对于我国这样各种灾害频发的国家，应该通过汶川地震灾后的重建工作总结出一套适合我国救灾的长效机制。特别希望通过新北川的建设在救灾重建方面为国家提供有益的经验。

在设计前期，我们编制了《北川新县城居住建筑设计导则》和《北川新县城居住建筑设计统一技术措施》。这两本技术文件在整个设计当中起到了非常重要的作用。《导则》和《技术措施》的编制依据《北川羌族自治县新县城灾后重建总体规划》，针对新县城受灾群众安置住房、拆迁村民安置住房、失地农民安置住房、廉租住房、租赁住房，针对选址与规划、建筑设计、结构设计、装修与设备、景观设计等方面进行了详细的规定和说明，保证质量和标准统一。

灾后重建需要居民的积极参与，不能单纯地给予，一个完整的家园不仅仅是靠建设一些房子来完成，其中的文化重建、精神重建将是更为艰巨的任务。

北川安居房工程已经建设完工并已入住，这对北川人民来说是一个新的开始，相信在此基础上新北川会发展得更加和谐、美好。

Affordable Housing Projects in the New County Seat of Beichuan

Liu Yanhui / Vice General Architect of China Architecture Design and Research Group

Housing construction is a key of post-quake reconstruction since it directly bears on the immediate interests of the general public, and is a reflection of our care to the quake-affiliated people and an important indicator measuring the achievements of post-quake construction of the new county seat of Beichuan. Thus the indemnificatory housing project is regarded as the top priority in the construction of the new county seat.

Residential buildings in the new county seat of Beichuan are composed of the resettlement housing for quake-affiliated people in Qushan Town of the original county seat and relocation housing for land-deprived farmers within the site boundary of the new county seat, and the total area is around 1 million m^2.

Residential building design mainly reflects the following six principles:

I. The principle of building a "safe, comfortable, vigorous, featured, civilized and harmonious" Beichuan county seat

The principle of building a "safe, comfortable, vigorous, featured, civilized and harmonious" Beichuan County represents the general requirements for the construction of Beichuan New County. As the basic criteria adopted in indemnificatory housing design, this policy is of vital importance to residential building design.

Based on living habits of Beichuan people, the structure of indemnificatory quarter adopts small neighborhood as the basic unit, with open space and coherent pedestrian system; the small size of neighborhood enhances the senses of belonging and intimate neighborhood relationships. Meanwhile, streets are arranged by utilizing natural landscape, so that more landscape resources can be shared by the residential quarter.

Continuance and inheritance of the Qiang culture is taken into full account in the architectural design of the indemnificatory housing project; in building style design, it extracts the Han-Qiang architectural symbols based on the "white stone worship" tradition of the Qiang people, and it creates a modern architectural style with strong Qiang characteristics by organically integrating traditional building materials with modern architectural structure; all these make the new county seat of Beichuan be of "distinctive features".

II. The principle of social justice

The indemnificatory housing project is designed for quake-affected people and relocated farmers, who concern much about the project as it is directly related to their own interests. Only by adopting the principle of social justice can a series of earthquake relief policies formulated by the Communist Party of China (CPC) and Chinese government be well implemented.

We have formulated different technical plans for projects of different features, and conducted in-depth investigation and research at the preliminary stage of planning to the satisfaction of the general public and the government, which plays an important role in realizing social justice and serves as a guidance of residential building design. By strictly following the government policy, we strictly checked our design for the specific house area and layout. We designed multiple house units of 50m^2, 70m^2, 90m^2, 105m^2 and 120m^2 in Er'ma Residential Quarter as per population composition feature of different families, with the error accuracy to 1m^2. In addition, storeroom, outdoor living balcony, open kitchen and toilet with windows are arranged according to the actual demands.

According to the resettlement method for the affordable housing in the new county seat of Beichuan, and in order to properly resettle residents as early as possible, the affordable houses are equipped with interior ceramic floor tiles and interior doors, kitchens and toilets are adopted with wall tiles, economic and water-saving sanitary wares and kitchen ranges, and interior painting are applied, so that the residents can move in without secondary decoration, and it reflects the design concept of comfortable and convenient building in details. The ground floor of the affordable housing is set with small shops, fully satisfying the demands of employment and operation and providing the residents with sufficient service facilities.

III. The principle of reflecting the "Qiang style and features"

Beichuan is the only Qiang autonomous county in China, and the idea of rebuilding a new county seat with Qiang style is widely accepted by the society. "Qiang style and features" not only shape a new image of Beichuan, but also is important element in creating tourism resources. As required in the general planning of Beichuan New County, since the new county seat is an area of Han-Qiang elements, the residence design gives fully consideration of regional culture characteristics, borrows ideas from the residential construction features, and combines with the local climatic characteristics and site conditions to make the residential buildings reflect traditional building features and regional culture.

Both flat roof and pitched roof are adopted in the affordable housing project, and the bottom floor uses external decorative face tiles while the top floor adopts wood color painting, reflecting regional architectural feature of a combination of wood and stone. In the meantime, the surrounding area is separated from the yard to enrich spatial organization, reduce cost and shorten construction period.

IV. The principle of "red flowers are relieved against green leaves"

The principle of "red flowers are relieved against green leaves" is adopted in the design of the new county seat of Beichuan where a number of buildings are built at the same time, and the aim is to further reflect the Qiang style and features. To be specific, for the whole Beichuan New County, public buildings should be "red flowers" while residential buildings be "green leaves"; for each residential quarter, buildings facing river and streets are "red flowers" while buildings at the center of cluster are "green leaves"; for residence, environment and landscape are "red flowers" while residential buildings are "green leaves". By rationally arranging the "red flowers" and "green leaves", the overall urban layout is well-structured, highlighting key landmark buildings while creating a quiet living atmosphere in a noisy neighborhood. It is also helpful to save investment and accelerate construction period.

Landscape design in the residential area also follows the principle of "red flowers are relieved against green leaves", and it focuses on "making fine design based on low-cost materials". For instance, we utilize permeable bricks of different colors to form Qiang embroidery pattern, which has very good effect. By considering lives of the Qiang people, and based on the regional culture, the landscape design advocates a blending of daily life and spiritual life closely related to the Qiang people by taking the Qiang tangible and intangible culture as well as folk custom as the main thread. Major Qiang festivals and celebrations in twelve months of a year are incorporated into the landscape design in the spirit of passing on and developing national culture, which are appreciated by the Qiang people as it reflects continuous regional culture in landscape design and enriches human landscape of the residential area.

V. The principle of energy saving, land saving and environmental protection

The principle of energy saving, land saving and environmental protection is a national policy advocated by the Chinese government, and it is reflected in the construction of the affordable housing project.

While ensuring comfortable living conditions and complete activity functions, the planning and design of the indemnificatory housing project increases the land use efficiency through rational planning layout and sound road organization. The cluster layout increases road density and facilitates travel of the residents. Green space landscape in the residential area fully utilizes the landscape belt around the new county and improves the greening quality. The indemnificatory houses are arranged as per national and local codes and standards, meeting the requirements of day lighting, ventilation and sunshine, and reflecting the solemnity of buildings built under the government's guidance.

The residential building design follows the requirements of Design Standard for Energy Efficiency of Residential Building of Sichuan Province and Evaluation Standard for Green Building with the aim of creating comfortable and pleasant living space. Guiding Principles for Designing Residence in the New County Seat of Beichuan and Uniform Technical Measures for Designing of Residence in the New County Seat of Beichuan were prepared at the preliminary design stage, which conform to the goal of building green architecture in terms of material selection, functional layout, structural nodes and construction control, so the whole construction process is under constant control.

The affordable housing project in the new county seat of Beichuan belongs to turnkey project, so the newly built residences meet basic living requirements, and the residents can move in after acceptance and delivery of the project. The design of the affordable housing adopts unified decoration standard to reduce decoration waste brought about by roughcast house, lower adverse impact of waste materials on environment, and effectively satisfy the construction schedule requirement made by the governments of different levels.

VI. The principle of establishing a national disaster relief system

China is a country frequently hit by various sudden disasters like earthquake, so we should establish a long-term disaster relief system suitable for disaster relief in China based on the reconstruction of Beichuan after the 5.12 Wenchuan earthquake, and we especially hope to offer useful experiences in disaster relief and post-disaster reconstruction.

At the preliminary design stage, we formulated Guiding Principles for Designing Residence in the new county seat of Beichuan and Uniform Technical Measures for Designing of Residence in the new county seat of Beichuan. These two technical documents are prepared based on the General Planning of Post-quake Reconstruction of the New County Seat of Beichuan Qiang Autonomous County and have played a very important role in the whole design process. They elaborate on site selection and planning, architectural design, structural design, decoration, mechanical and landscape design by analyzing different types of housing, such as resettlement housing for the quake-affiliated people in the new county, resettlement housing for the relocated farmers, resettlement housing for the land-deprived peasants, low-rent housing and rental housing to make project quality and standard be consistent.

The Beichuan people are expected to actively participate in the post-quake reconstruction instead of simply receiving. A complete homeland can not be realized only by building several houses; rather, cultural and spiritual reconstruction will be a more arduous task.

With the completion and delivery of Beichuan affordable housing project, Beichuan people embrace a new beginning, and we hope that the new county seat of Beichuan will become more harmonious and beautiful in the future.

尔玛小区 Er'ma Residential Quarter

援　建：山东省济南市 青岛市 淄博市 烟台市
　　　　潍坊市 济宁市 威海市 临沂市
方　案：中国建筑设计研究院
　　　　中国城市规划设计研究院
施工图：中国建筑设计研究院
　　　　山东同圆设计集团有限公司
　　　　中国航天建筑设计研究院
　　　　青岛市建筑设计研究院有限公司
　　　　青岛理工大学建筑设计研究院
　　　　淄博四新建筑设计有限公司
　　　　烟台市建筑设计研究股份有限公司
　　　　潍坊市建筑设计研究院有限责任公司
　　　　济宁市建筑设计研究院
　　　　威海市建筑设计院有限公司
　　　　临沂市规划建筑设计研究院
施　工：山东平安建设集团有限公司
　　　　济南二建集团工程有限公司
　　　　青岛一建集团有限公司
　　　　青建集团股份有限公司
　　　　莱西市建筑总公司
　　　　山东金城建工有限公司
　　　　山东金塔建设有限公司
　　　　山东新城建设有限公司
　　　　烟建集团有限公司
　　　　潍坊昌大建设集团有限公司
　　　　潍坊市三建集团有限公司
　　　　青州市第一建筑工程有限公司
　　　　山东圣大建设集团有限公司
　　　　山东鸿顺建设集团有限公司
　　　　威海建设集团股份有限公司
　　　　山东华鲁建安集团有限公司

Aided by:
Jinan, Qingdao, Zibo, Yantai, Weifang, Jining, Weihai and Linyi of Shandong Province
Scheme design:
China Architecture Design & Research Group
China Academy of Urban Planning & Design
Construction drawings:
China Architecture Design & Research Group
Shandong Tongyuan Design Group Co., Ltd.
China Space Civil & Building Design & Research Institute
Qingdao Architectural Design & Research Institute
Science and Engineering University Building Design Institute for Research of Qingdao
Zibo Sixin Architectural Design Co., Ltd.
Yantai Architectural Design and Research Co., Ltd.
Weifang Institute of Architectural Design
Jining Architecture Design & Research Institute
Weihai Architectural Design Institute Co., Ltd.
Planning & Architecture Design Institute of Linyi
Contractors:
Shandong Ping'an Construction Group Co., Ltd.
Jinan No.2 Construction Engineering Company
Qingdao No.1 Construction Group Co., Ltd.
Qingjian Group Co., Ltd.
General Construction Company of Laixi
Shandong Jincheng Construction Engineering Co., Ltd.
Shandong Jinta Construction Co., Ltd.
Shandong Xincheng Construction Co., Ltd.
Yanjian Group Co., Ltd., Shandong Hualu Jian'an Group
Weifang Changda Construction Group Co., Ltd.
Weifang No. 3 Construction Group Co. Ltd.
Qingzhou No.1 Construction Engineering Co., Ltd.
Shandong Shengda Construction Group Co., Ltd.
Shandong Hongshun Construction Group Co., Ltd.
Weihai Construction Group Co., Ltd.
Shandong Hualu Jian'An Group Co., Ltd.

用地面积：284 200m^2
建筑面积：421 500m^2
容纳户数：3 638户

Site area: 284,200m^2
Building area: 421,500m^2
Capacity: 3,638 suites

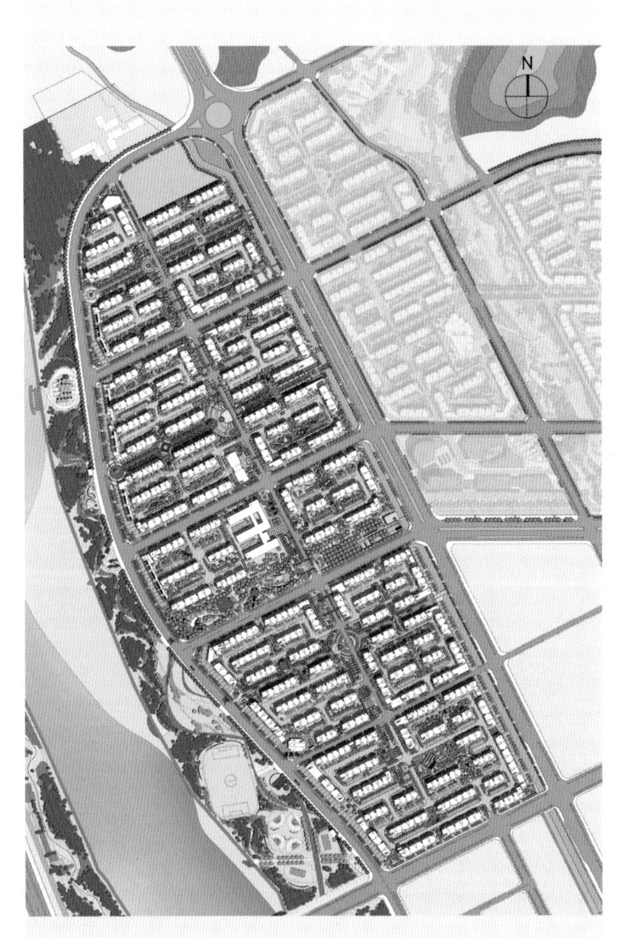

总平面图 Site plan

尔玛小区是北川新县城安居房的重要组成部分，用以满足部分原北川县受灾群众的居住需求，是新县城的先期启动项目，位于新县城西北部，其规划和建筑都充分体现了羌族文化特点。

规划以小街坊组织空间，实现了开放的街区和连贯的步行系统，强调了居住区的均好性和归属感，易于促成邻里间的亲密关系，较密的道路网也有利于分担交通流量。住宅间的带状绿地巧妙将小区西侧城市公园的景观引入居住区内，将原黄土镇的部分民居、石碑和石桥保留为景观设施，并安排了羌族跳锅庄舞的小广场。

安居房的户型依据安置住宅的面积规定设计，满足自然通风、采光等绿色标准，注重与北川当地经济发展水平和居民生活习惯相适应。立面设计尊重地域文化特色，提取当地民居汉羌建筑符号，吸收羌族〝白石〞崇拜的传统，底层为外装饰面砖，顶层为木色涂料，屋顶采用平坡结合的形式，充分体现当地民居的地域特征。

组团入口结合羌族元素，体现民族特色 The community entrance is designed with distinctive Qiang feature to emphasize the ethic character of the project.

The Er'ma Residential Quarter is one important component of the affordable housing project which is located in the northwest of the new county seat. As one of early initiated projects, this quarter is built for meeting the living demands of parts of the quake-afflicted people of the former county seat, and its planning and architectural style fully reflect the Qiang ethnic culture.

In a neighborhood organization, the planning realizes open block streets and coherent walking flow, and focuses on the harmony and sense of belonging of the residential community, which facilitates intimacy of the neighborhood and less burden on the traffic network. Besides, the greenbelts in between the quarter skillfully introduce the park landscape on the west side of the community into its compound, with parts of dwelling house, stone tablet and stone bridge of the original Huangtu Town being reserved as landscape facilities, and small squares for the Qiang Guozhuang dance are also arranged.

The house types are designed based on the area provisions of resettlement, which satisfy the green standards of natural ventilation and day lighting, and are compatible with the local economic development level and living habits. In addition, the façade design gives full respect to the local culture, refines the Han-Qiang architectural symbols of the local dwelling houses, and incorporates the Qiang tradition of "worship of white stone" by applying exterior decorative face tiles for the bottom, timber-color coating for the top and a combination of flat and pitched roof to completely reflect the regional features of the local dwelling houses.

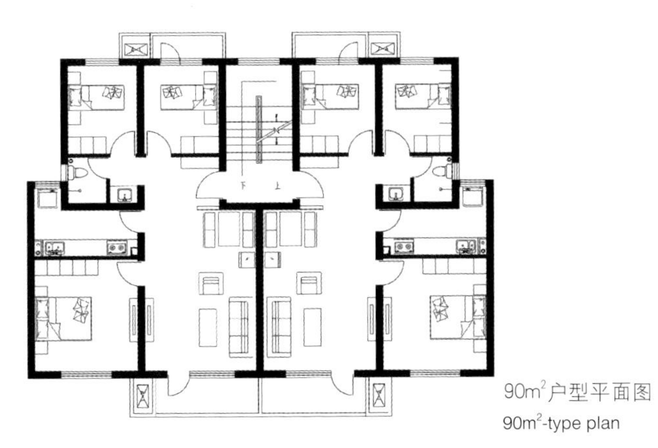

90m²户型平面图

90m²-type plan

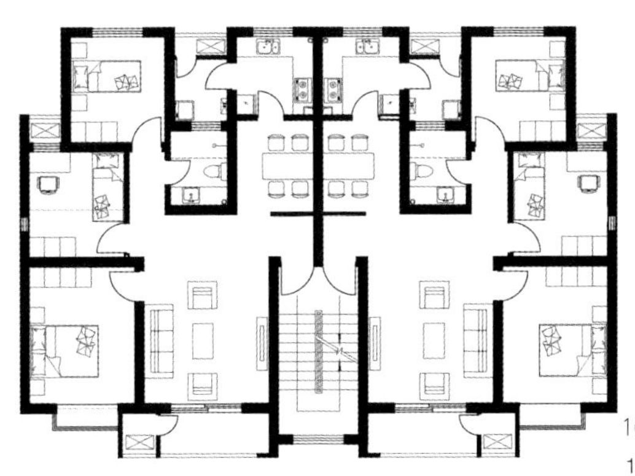

105m²户型平面图

105m²-type plan

羌族传统石塔"拉克西"成为组团广场的中心元素 The Qiang traditional stone pagoda "Lakexi" is the central element of the residential quarter square.

小区内安放了12块代表羌族传统节日的石头，作为各处景观的标志 As symbols of landscape, twelve stones that represent the Qiang traditional festivals are placed in the residential quarter.

景观主轴线的设计将羌绣图案融入铺地 The Qiang ethnic embroidery patterns are harmoniously incorporated in the pavement by the design of the main axis of landscape.

小区内为回族居民特别建设的清真寺

A mosque is specially built for Muslim residents.

永昌幼儿园北园 Yongchang Kindergarten (North Part)

援　建：山东省枣庄市

方　案：中国建筑设计研究院

施工图：枣庄市建筑设计研究院

施　工：滕州市建筑安装工程集团公司

用地面积：7 051m²

建筑面积：4 810m²

建筑规模：18班 540人

Aided by:
Zaozhuang, Shandong Province
Scheme design:
China Architecture Design & Research Group
Construction drawings:
Zaozhuang Institute of Architectural Design and Research
Contractor:
Tengzhou Construction and Installation Engineering Group Corporation

Site area: 7,051m²
Building area: 4,810m²
Capacity: 18 classes with 540 children in total

永昌幼儿园分为南、北两个部分。北园位于尔玛小区内，采用院落围合式布局，主入口设于用地东侧，一条贯穿南北的功能廊不仅容纳了办公用房，更将其西的三组儿童活动单元和位于用地南、北侧的音体教室、后勤厨房联系起来。儿童活动单元采用模块化设计，由活动室、卧室、衣帽间、卫生间四部分组成。每6个单元为一组，排列于用地西侧，为幼儿提供了便捷、舒适的室外活动空间。建筑立面变化丰富，色彩组合鲜艳明快，充分考虑了幼儿的心理和尺度需求。

Yongchang Kindergarten consists of north and south parts. The north part is located in the Er'ma Residential Quarter, where a courtyard layout is enclosed with the main entrance on the east side of the site. A functional gallery running through north and south accommodates all offices and connects three groups of children activity units on the west, with the music and sports room on the south and the kitchen on the north. The children activity units use a modular design with four parts including the activity room, bedroom, cloakroom and toilet. Aligning along the west side of the project site, each group of six units provides convenient and comfortable outdoor activity space for children. In addition, the building design also gives full consideration to children's psychological demand and dimension requirement by using various façade changes and bright combinations of colors.

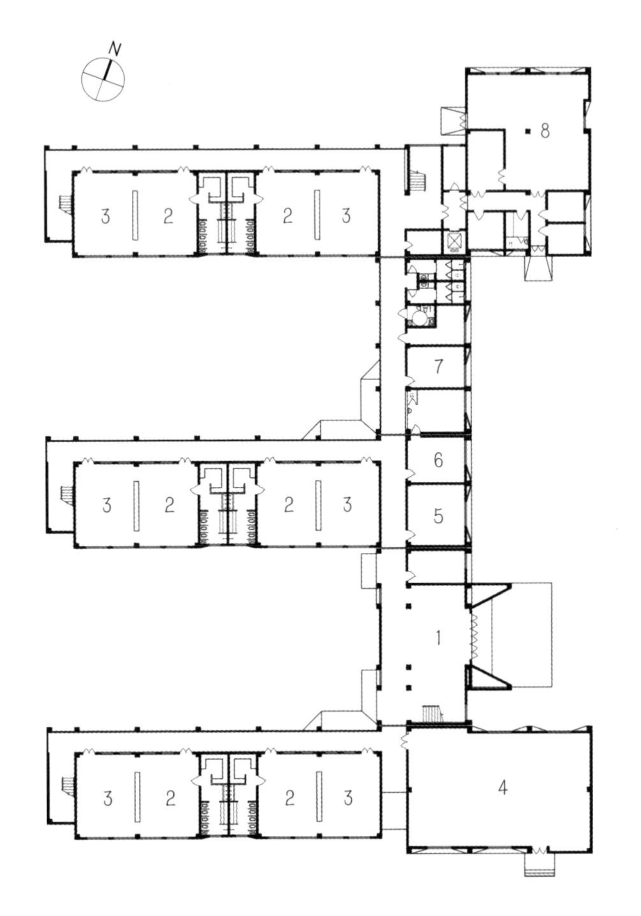

1. 门厅 foyer
2. 活动室 activity room
3. 卧室 bedroom
4. 音体活动室 music and sports room
5. 晨检、接待 morning examination and reception
6. 保健室 health care room
7. 办公室 office
8. 厨房 kitchen

首层平面图 Ground floor plan

门斗和各色窗洞为孩子们提供羌族民居空间体验的朦胧印象　Foyer and colorful window walls give children a hazy impression of the spatial experience of the Qiang folk house.

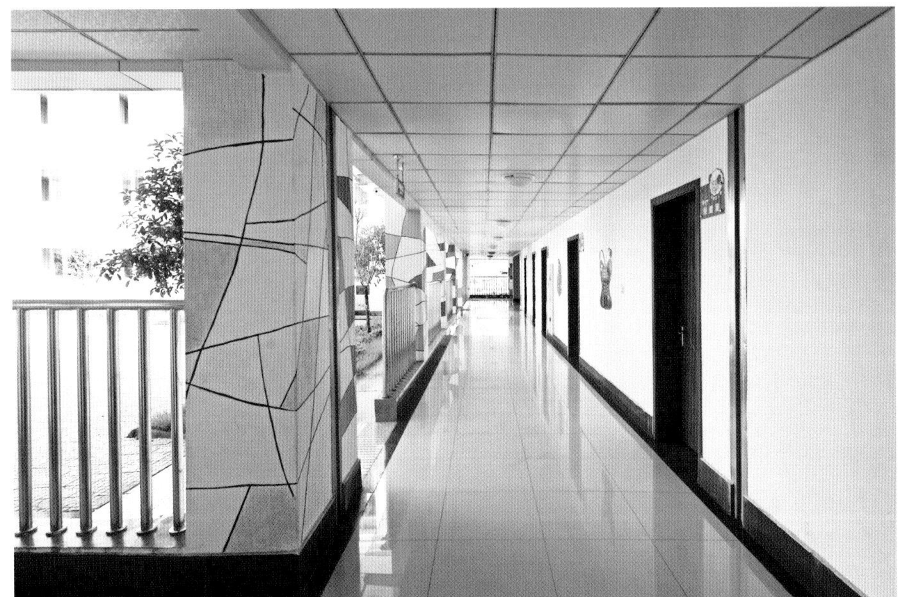

半外廊、露台和庭院组成丰富的室外活动空间，既适应当地气候，也为孩子们提供多种空间选择

Corridors, terraces and courtyards form rich outdoor activity space, which not only adapts to the local climate but also provide multiple spaces for children to play.

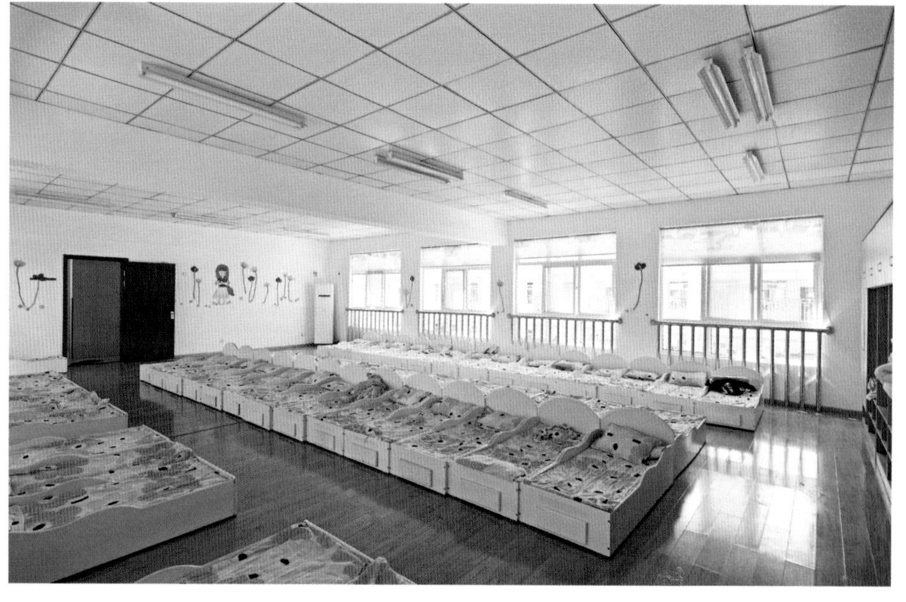

新川小区 Xinchuan Residential Quarter

承　建：山东省济宁市、威海市
　　　　德州市、菏泽市
方　案：中国建筑设计研究院
　　　　中国城市规划设计研究院
施工图：北京国城建筑设计公司
　　　　中国建筑设计研究院
施　工：山东圣大建设集团有限公司
　　　　山东宁建建设集团有限公司
　　　　山东鸿顺建设集团有限公司
　　　　威海建设集团股份有限公司
　　　　山东荣城建筑集团有限公司
　　　　山东德建集团
　　　　山东德兴建设集团有限公司
　　　　德州天元集团有限公司
　　　　山东菏建建筑集团有限公司

Aided by:
Jining, Weihai, Dezhou and Heze of Shandong Province
Scheme design:
China Architecture Design & Research Group
China Academy of Urban Planning & Design
Construction drawings:
Beijing Guocheng Architectural Design Company
China Architecture Design & Research Group
Contractors:
Shandong Shengda Construction Group Co., Ltd.
Shandong Ningjian Construction Group Co., Ltd.
Shandong Hongshun Construction Group Co., Ltd.
Weihai Construction Group Co., Ltd.
Shandong Rongcheng Construction Group Co., Ltd.
Shandong Dejian Group Co., Ltd.
Shandong Dexing Construction Group Co., Ltd.
Dezhou Tianyuan Group Co., Ltd.
Shangdong Hejian Construction Group Co., Ltd.

新川小区是为原黄十镇居民提供的拆迁安置住宅，位于新县城南端。设计延续街坊式格局，不仅有利于邻里交往和社区安全，也便于营造连续的沿街界面。组团划分兼顾原有四个村的居民组织，相互独立又彼此为邻，通过公共设施、景观资源、环境品质和户型标准的均匀设置，确保社会公平。考虑到社区管理和物业维持的承受能力，绿地尽量外置于城市公共范畴的外围滨河带和休闲公园内，通过延续永昌河和安昌河的景观脉络，实现景观资源的“生长”。公共服务设施既有片区集中设置的幼儿园、市场、派出所等，也有沿社区周边布置的商业，为居民提供本地就业机会。户型设计上尊重地域生活习惯，尽可能多地设置储藏空间，实现明厨明卫。建筑外观充分尊重地域特色，借鉴当地民居特点，从材质和颜色上体现木石结合的地域建筑特征。

Situated on the south end of the new county seat, the Xinchuan Residential Quarter aims to provide resettlement for residents of the former Huangtu town. Following a neighborhood-style layout, the design not only promotes close neighborhood and community security, but also facilitates a continuous interface along the streets. The community is divided into four groups by following the four original villages, keeping the groups independent and intimate with an even arrangement of public facilities, landscape resources, environment quality and house-type standard. Based on the capacity of community management and property maintenance, the design places the greening space out of the quarter to the riverside and leisure park where the landscape of the Yongchang River and Anchang River is extended to realize “growth” of landscape resources. The public service facilities include kindergarten, market and police station within the community, and also incorporate shops around the community to create more employment opportunities for local residents. The house types are designed to respect the local living habit, leave more space for storerooms, and realize bright kitchens and toilets. Moreover, the building façade shows full respect to the local characteristics by applying a combination of timber and stone in materials and colors.

用地面积：218 700m^2
建筑面积：325 884m^2
容纳户数：3 161户

Site area: 218,700m^2
Building area: 325,884m^2
Capacity: 3,161 suites

总平面图　Site plan

建筑的民族特色主要通过底层商业的部分体现

The ethnic features of the buildings are reflected mainly in the shops on the ground floor.

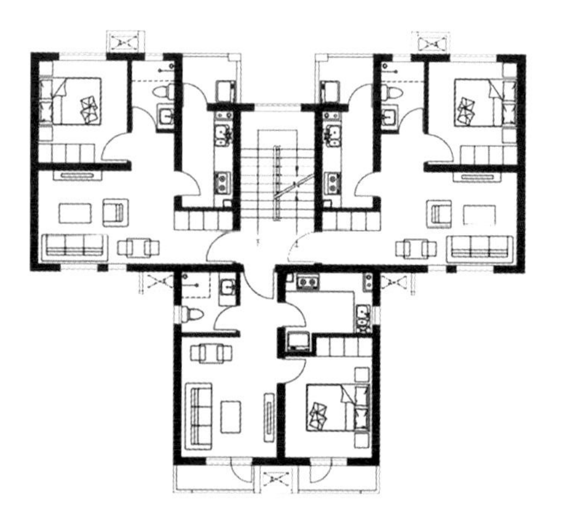

48m²户型平面图　48m²-type plan

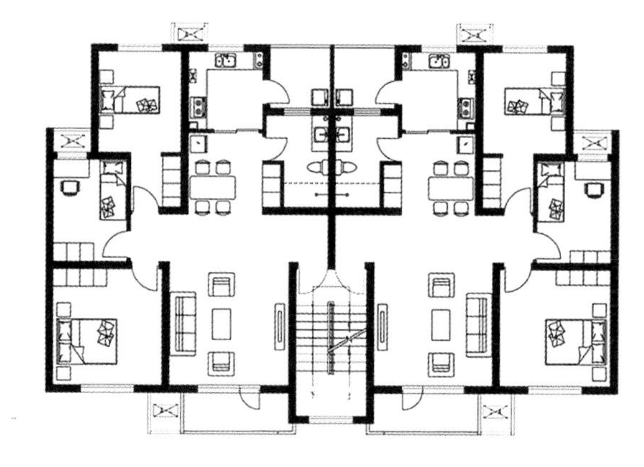

105m²户型平面图　105m²-type plan

永昌幼儿园南园 Yongchang Kindergarten (South Part)

援　建：山东省枣庄市

方　案：中国建筑设计研究院

施工图：北京国城建筑设计公司

施　工：滕州市建筑安装工程集团公司

用地面积：7 153m²

建筑面积：3 373m²

建筑规模：12班 360人

Aided by:
Zaozhuang, Shandong Province
Scheme design:
China Architecture Design & Research Group
Construction drawings:
Beijing Guocheng Architectural Design Company
Contractor:
Tengzhou Construction and Installation Engineering Group Corporation

Site area: 7,153m²
Building area: 3,373m²
Capacity: 12 classes with 360 children in total

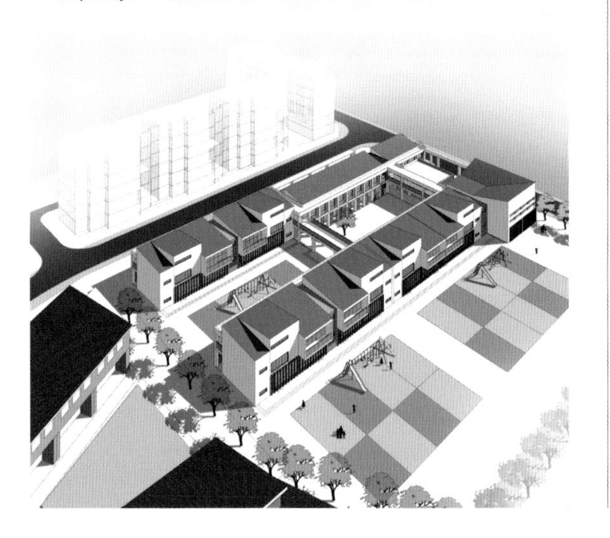

永昌幼儿园南园位于新川小区内，永昌河景观带西侧。两条平行布置的儿童活动单元组成两条两层高的建筑体形，平行布置于用地中，围合形成的院落空间不仅为幼儿提供活动场地，也丰富了室内外的空间感受。主入口位于用地东侧，北为医务室和后勤厨房，南为音体教室。儿童活动单元同样采用模块化设计，由活动室、卧室、衣帽间、卫生间四部分组成。立面设计强调白色涂料墙面与木色装饰的对比，轻重对比，充分尊重永昌河景观带，并从川西民居中提取特征，反映地域特色，立面丰富的变化则为建筑增添了童趣。

The South Part of Yongchang Kindergarten is located in the Xinchuan Residential Quarter along the west side of the landscaped green belt along Yongchang River. Two paralleling activity units for children form two building masses of two stories. Such an enclosed courtyard not only provides activity space for children, but also enriches outdoor and indoor spatial experiences. The main entrance of the kindergarten is set on the east side of the site; on the north are health care room and kitchen, and on the south is a music and sports room. Same as the North Part, the children activity units adopt a modular design with four parts including the activity room, bedroom, cloakroom and toilet. Moreover, the façade design lays emphasis on contrasts of the white-coated wall and timber decoration, and of lightness and heaviness. The design shows full respect to the landscaped green belt along Yongchang River, refines features of the local dwelling houses, and the rich changes on the façade endow joy for children.

永昌小区 Yongchang Residential Quarter

承　建：山东省烟台市、日照市、聊城市

方　案：中国建筑设计研究院

中国城市规划设计研究院

施工图：北京国城建筑设计公司

中国城市规划设计研究院

施　工：烟建集团有限公司

烟台新世纪工程项目管理咨询有限公司

山东泰和建设管理有限公司

山东日照建设集团有限公司

山东锦华建设集团有限公司

山东聊建集团总公司

Aided by:

Yantai, Rizhao and Liaocheng, Shandong Province

Scheme design:

China Architecture Design & Research Group

China Academy of Urban Planning & Design

Construction drawings:

Beijing Guocheng Architectural Design Company

China Academy of Urban Planning & Design

Contractors:

Yanjian Group Co., Ltd.

Yanjian New Century Engineering Project Manage & Consulting Co., Ltd.

Shandong Taihe Construction Supervision Co., Ltd.

Shandong Rizhao Construction Group Co., Ltd.

Shandong Jinhua Construction Group Co., Ltd.

Shandong Liaocheng Construction Group Co., Ltd.

永昌小区同样是为新县城原址居民提供的拆迁安置区，位于新县城中轴线东端，靠近山体，地理位置相对独立。规划布局延续了城市中轴线在空间和视觉上的连续感。设计将规模商业集中布置，与住宅相对分开，减少商业人流对住宅的干扰。住宅组团采用院落式围合布局，增强了组团的归属感。内部交通采用“机非分流”交通系统，沿组团内规划环形车行系统，有效减少车辆对内部环境的影响，为创造祥和宁静的内部空间创造有利条件。为衬托自然山体，建筑形式较为简洁，主要通过天际线的起伏和体块的错落体现对传统羌族村寨的呼应。小区北部采用坡屋顶，体现地域特色，南部则强调与工业区的联系，采用更具时代感的平屋顶。

The Yongchang Residential Quarter is also a resettlement for the former resident. Located on the east end of the central axis of the new county seat, it is close to mountains and enjoys an independent geographical position. The planning layout extends the central axis in terms of spatial and visual senses. In this design, the large-scale business area is separated from the residence to reduce disturbance of human flow from the business side. Building clusters in the quarter apply an enclosed layout of courtyard to enhance a sense of belonging. The internal traffic adopts a “separation of vehicle and non-vehicle” with circular vehicle paths around the community, which can effectively minimize influences of vehicles on the internal environment, and create favorable conditions for building peaceful internal space. The appearance of buildings is relatively concise, and responds to the traditional Qiang villages mainly through a fluctuant skyline and intersperse building mass. The north of the residential quarter applies pitched roofs to reflect the local features, while the south side uses modern flat roofs to harmonize its style with the industrial park.

用地面积：171 300m²　　Site area: 171,300m²

建筑面积：251 419m²　　Building area: 251,419m²

容纳户数：2 344户　　Capacity: 2,344 suites

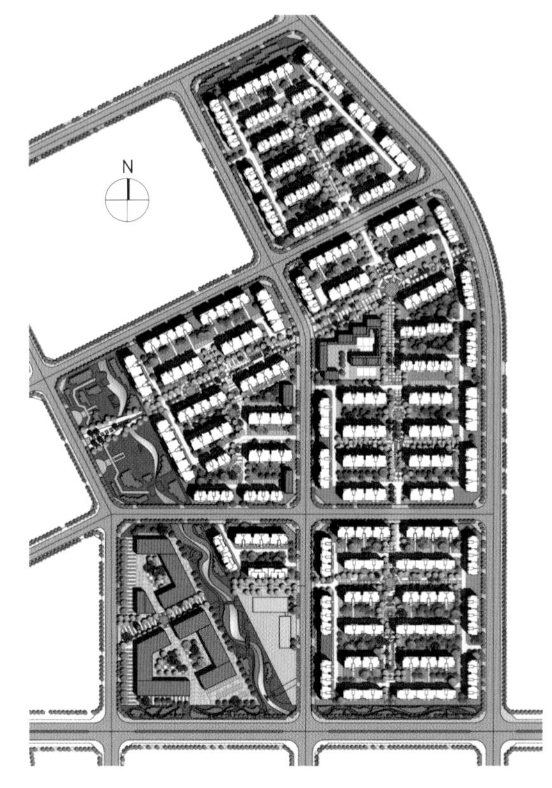

总平面图 Site plan

30

禹龙小区 Yulong Residential Quarter

承　建：山东省淄博市、潍坊市
方　案：中国城市规划设计研究院
　　　　中国建筑设计研究院
施工图：北京国城建筑设计公司
　　　　青岛理工大学建筑设计研究院
施　工：山东高阳建设有限公司
　　　　山东天齐置业集团股份有限公司
　　　　山东金塔建设有限公司
　　　　山东新城建设有限公司
　　　　潍坊昌大建设集团有限公司
　　　　潍坊市三建集团有限公司
　　　　青州市第一建筑工程有限公司

Aided by:
Zibo and Weifang, Shandong Province
Scheme design:
China Academy of Urban Planning & Design
Science and Engineering University Building Design Institute for Research of Qingdao
China Architecture Design & Research Group
Construction drawings:
Beijing Guocheng Architectural Design Company
Science and Engineering University Building Design Institute for Research of Qingdao
Contractors:
Shandong Gaoyang Construction Co., Ltd.
Shandong Tanki Industry & Commerce (Group) Co., Ltd.
Shandong Jinta Construction Co., Ltd.
Shandong Xincheng Construction Co., Ltd.
Weifang Changda Construction Group Co., Ltd.
Weifang No. 3 Construction Group
Qingzhou No. 1 Construction Engineering Co., Ltd.

禹龙小区位于新县城北部，北为自然山麓，西邻尔玛小区。其居民既有原北川县的受灾群众，也有当地需要拆迁安置的群众，设计因而部分延续了尔玛小区对羌族风貌的传承，并适当增强了建筑的现代特性。永昌河景观带从小区中穿过，形成滨水公园，结合水系安排的带状绿地则进一步将景观延伸到组团内部，同时延续了尔玛小区的景观轴线，增强了二者空间上的对应和联系。建筑在形式、色彩和材料上与尔玛小区保持协调一致，屋顶形式则更为简练地呈现了羌族村寨的风貌。

The Yulong Residential Quarter is located on the north side of the new county seat, and it borders the natural mountain root on the north and adjacent to the Er'ma Residential Quarter on the west. It is home to the quake-affiliated people who once lived in the old Beichuan county seat before the earthquake and relocated local residents. The design carries on the Qiang styles and features similar to the design for Erma Residential Quarter, and properly enhances the modern feature of the building. The landsacaped green belt along Yongchang River runs through it and a riverside park is thus formed. The green belt arranged in combination with a water system further extends the landscape into the interior of the community, and it shares the landscape axis of Er'ma Residential Quarter, thus enhancing spatial communication of the two. It shares the same form, color and materials with Er'ma Residential Quarter, and its roof demonstrates the styles of the Qiang villages in a more concise form.

总平面图 Site plan

用地面积：277 700m² 　Site area: 277,700m²
建筑面积：289 269m² 　Building area: 289,269m²
容纳户数：2 466户 　Capacity: 2,466 suites

公共服务设施概况

Public Service Facilities

自2009年3月四川省人民政府正式批复《北川羌族自治县新县城灾后重建总体规划》后，中规院驻北川新县城现场工作组历时2个多月，完成北川新县城绝大部分公共服务设施建设标准和具体建设内容的核定，并确定了各个项目的选址和用地规模。为此后各设施项目的建筑设计方案招标、山东援川办公室确定援建计划奠定了基础。

北川新县城公共服务设施建设标准及具体建设内容的确定是一项政策性强、涉及各方利益、民众高度关注的工作。本着节约用地、高效使用援建资金、综合利用重建设施和降低建成后的日常运营维护成本等原则，对《汶川特大地震灾后重建总体规划》中涉及北川县城的重建项目220余项进行了深入细致的梳理，最终确定7大类、66个重建项目纳入新县城山东援建和地方重建项目库，包含了教育科研、社会福利、体育、文化娱乐、行政办公、医疗卫生等设施。最终审核确认进入新县城的项目中，仅公益性服务服务设施一项，取消和整合了一大批项目，建筑规模比原先项目库的压缩超过4万平方米，为灾后重建节约资金超过2亿元，为新县城今后发展保留了大量可开发土地。（殷会良 执笔）

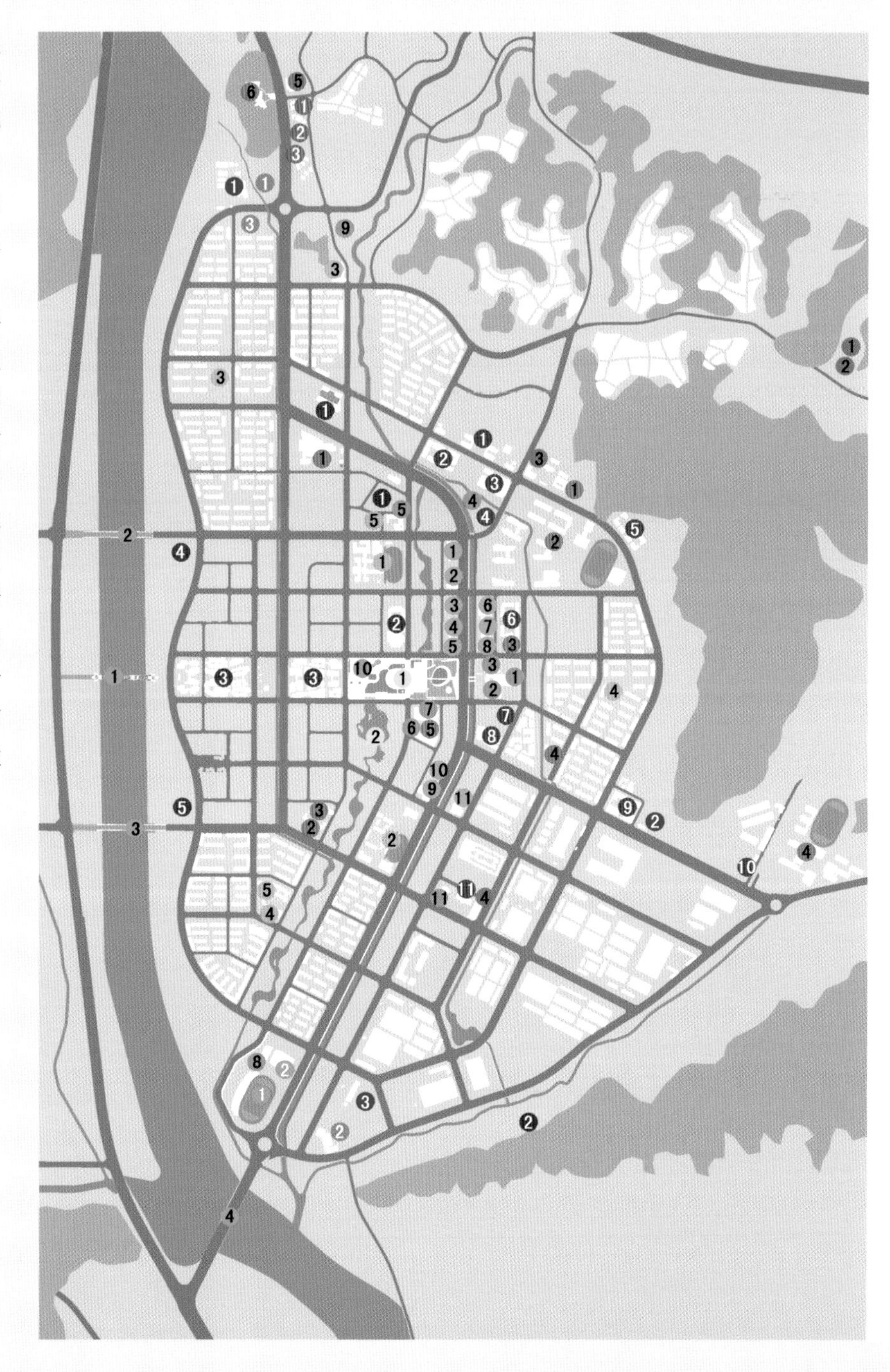

Since the General Planning of Post-quake Reconstruction of the New County Seat of Beichuan Qiang Autonomous County was approved officially by Sichuan Provincial People's Government in March 2009, the Front-Line Headquarters for New Beichuan New County Seat On-site Task Force of CAUPD took two months to check the construction standards and specific construction contents of the majority of public service facilities in the new county seat of Beichuan, and completed site selection for all engineering works and confirmed their respective land sizes, thus laying foundation for the architectural design scheme bidding and aid-construction plan confirmation of Headquarters of Projects aided by Shandong Province.

The confirmation of the construction standards and specific construction contents of the public service facilities in the new county seat of Beichuan is a policy-oriented task which relates to the interests of the stakeholders and is highly concerned by the general public. In the light of the principles of saving land, effectively utilizing the grant of aid, comprehensively using reconstructed facilities and reducing daily operation and maintenance costs after completion, we studied on over 220 projects related to the construction of the new county seat of Beichuan in the *General Planning on the Reconstruction after the 5.12 Wenchuan Earthquake,* and 66 reconstruction projects in 7 categories are finally incorporated into the databases of projects aided by Shandong Province and local reconstruction projects, including facilities for education and scientific research, social welfare, sports, culture and entertainment, office and health care. Among the projects reviewed and finally included in the construction of the new county seat of Beichuan, and for the item of public service facilities only, a host of projects were canceled or integrated, thus the total building scale is 40,000m^2 less than that included in the original project database, saving reconstruction funds by over 200 million RMB, and retaining a great quantity of land for future development of the new county. (by Yin Huiliang)

永昌小学 Yongchang Primary School

援　建：山东省淄博市

方　案：中国建筑设计研究院

施工图：淄博四新建筑设计有限公司

施　工：山东齐泰实业集团股份有限公司

用地面积：34 000m²

建筑面积：14 400m²

建筑规模：36班　1620人

Aided by:
Zibo, Shandong Province
Scheme design:
China Architecture Design & Research Group
Construction drawings:
Zibo Sixin Architectural Design Co., Ltd.
Contractor:
Shandong Qitai Industrial Group Co., Ltd.

Site area: 34,000m²
Building area: 14,400m²
Capacity: 36 classes with 1,620 students in total

永昌小学位于新县城中心区北侧，东邻永昌河景观带。校园由教学楼、学生宿舍、学生食堂、设备用房等部分组成。设计的目的是创建一个结构安全与心理安全并重的校园，注重快速疏散，建筑层数尽量控制在3层以内，利用不同高度的屋顶平台、连廊、体育场的连接，提供更多的疏散路径。高效、多元的联系空间既营造出一个丰富立体的校园空间，为学生们创造了多样的交往空间和活跃、亲切的学习环境，也是对羌族聚落层叠错落的屋顶平台意象的表达，羌族民居的坡屋顶、碉楼、过街楼、干栏式连廊、花窗、石片墙等元素也同样在建筑上有所反映。

Yongchang Primary School is located on the north of the new county seat. It is composed of classroom building, dormitory buildings, dining hall and equipment room. The design not only emphasizes the structural safety, but also cares the psychological security. Fast evacuation function is stressed in the design, and the number of building floor is controlled within three. More evacuation paths are provided by connecting roof platforms, corridor and stadium at different heights. The efficient and diversified communication space makes it a diversified and lively environment for study. The design also intends to present the image of scattered roof platform of the Qiang village. Elements like pitched roof, turret, arcade, corridor, patterned window and stone wall are well applied in the whole project.

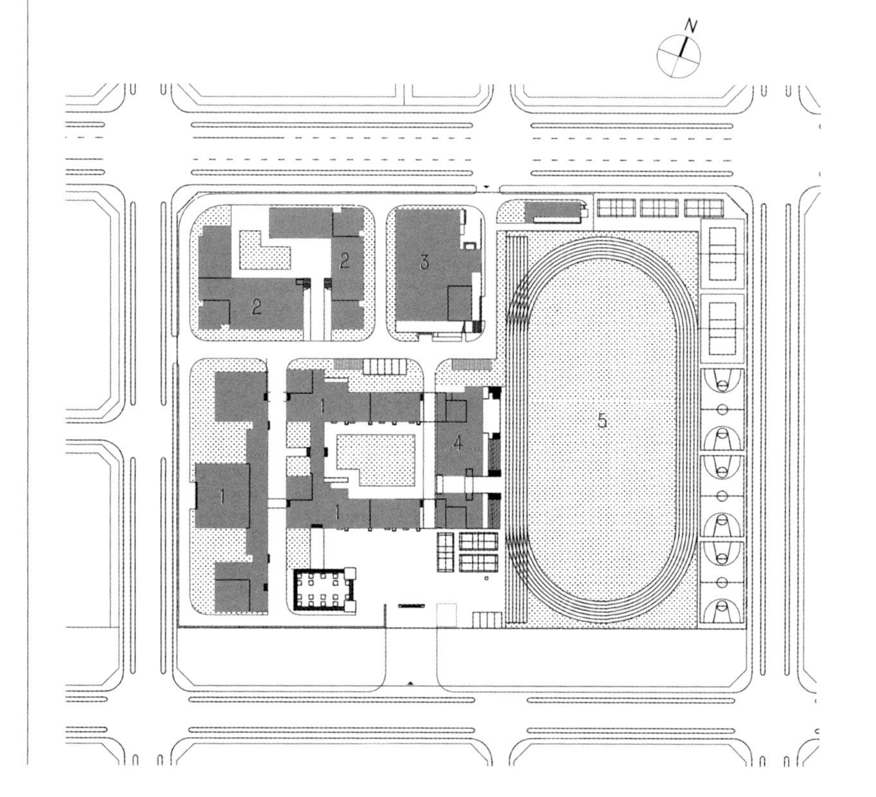

1 教学楼 classroom building
2 宿舍楼 dormitory building
3 食堂 dining hall
4 体育教室 gymnasium
5 运动场 sports ground

总平面图 Site plan

地面层、教学楼屋顶层，以及各功能部分之间的联系平台，共同营造出一个丰富的立体校园空间

The ground floor, the roof level of the classroom building and platforms for connecting each functional part create the campus space in three dimensions.

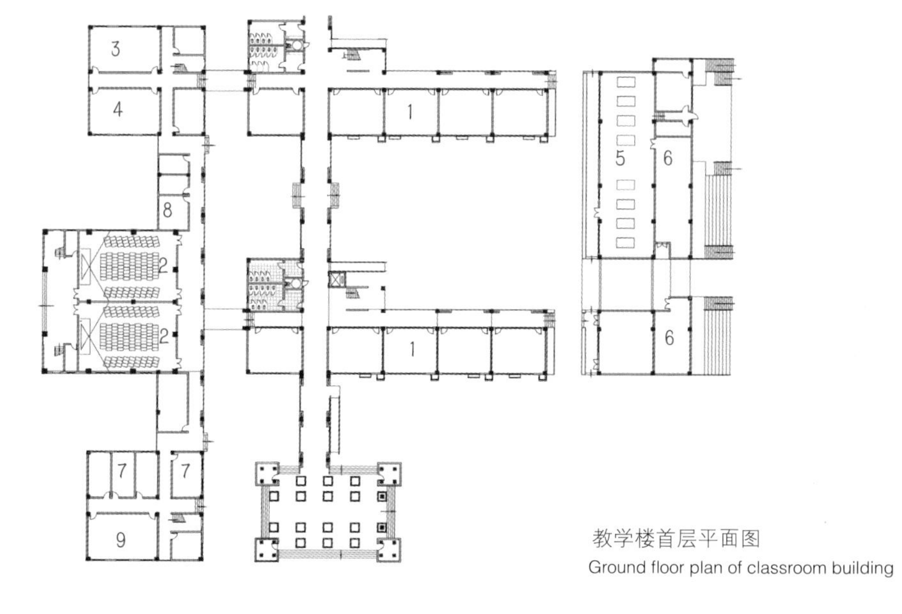

教学楼首层平面图
Ground floor plan of classroom building

1 普通教室 typical classroom
2 合班教室 large classroom
3 劳技教室 labour classroom
4 自然教室 science classroom
5 乒乓球训练室 table tennis training room
6 器材室 sports equipment room
7 办公室 office
8 保健室 health care room
9 仓库 storeroom

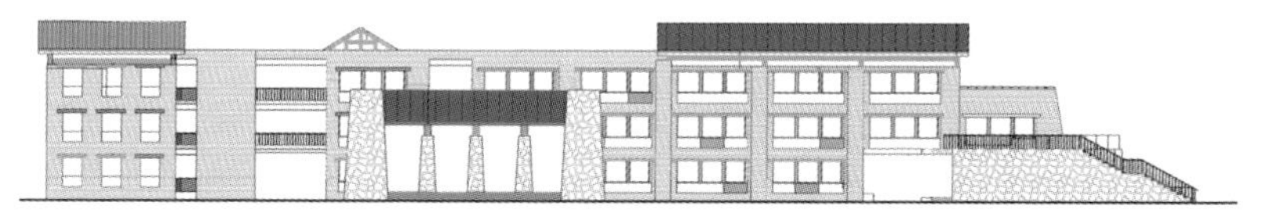

教学楼南立面图 South elevation of classroom building

立足岗位做贡献 创先争优当模范
各位领导和老师莅临我校指导工作!
山东淄博援建永昌小学

永昌中学 Yongchang Middle School

援　建：山东省德州市
设　计：德州市建筑规划勘察设计研究院
施　工：山东德建集团有限公司

用地面积：32 800m²
建筑面积：14 863m²
建筑规模：36班　1620人

Aided by:
Dezhou, Shandong province
Designer:
Dezhou Architecture Planning Survey and Design Institute
Contractor:
Shandong Dejian Group Co., Ltd

Site area: 32,800m²
Building area: 14,863m²
Capacity: 36 classes with 1,620 students in total

永昌中学位于新县城南部，东为城市干道永昌大道，西为永昌河景观走廊。设计既通过方整的校园空间为道路营造城市界面，也将景观引入校园。体育运动区位于用地东侧，遮挡了主干路噪声对教学区的影响。学生宿舍及食堂则在地块南侧形成完整的城市界面。建筑体量遵循方格网布置，又引入丰富的内部变化。入口广场以办公楼作为主要建筑体量，模仿了羌族村寨以碉楼为中心的空间序列。组群单元形成基本院落空间，以架空廊道与外部联系，形成必要的灰空间。屋面平坡结合，形成丰富的轮廓线。

Yongchang Middle School is located on the south side of the new county seat, between the Yongchang Road and the landscaped green belt along Yongchang River. Its layout creates an urban interface for roads through the regular campus space while introduces landscape into the campus. The sports area is on the east side, sheltering the impact of noise of the main roads on the teaching area. The student dormitory and dining hall in the south form a complete urban interface. The building mass is arranged with a grid mesh and introducing rich internal changes. The office building facing the main entrance imitates the spatial status of the turret in Qiang village. The complex unit forms basic yard space and connects with the outside with elevated corridor to form necessary grey space. Flat and pitched roofs are used to generate rich contour line.

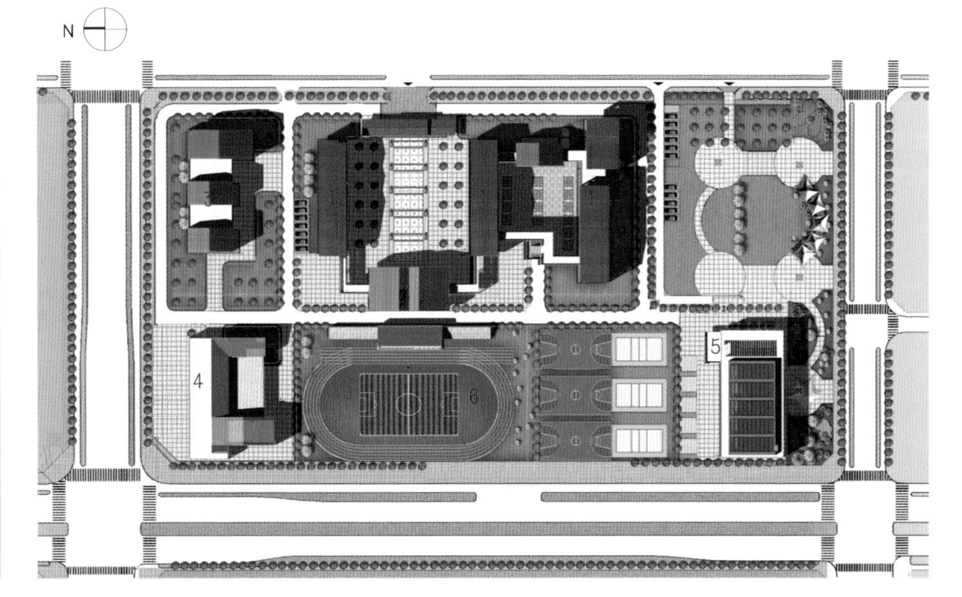

1 教学楼 classroom building
2 办公楼 office building
3 宿舍楼 dormitory building
4 食堂 dining hall
5 风雨操场 gymnasium
6 运动场 sports ground

总平面图 Site plan

折板式的体形单元间穿插了室外及半室外空间，形成丰富的空间体验，也获得生动的建筑轮廓线

Folded-plate units intersperse between the outdoor and semi-outdoor space, forming abundant spatial experiences and acquiring a vigorous contour of the buildings.

院落式的布置，在教学区形成独立而便于交流的室外空间

The courtyard layout forms separate outdoor space from the classroom area that is convenient for communication.

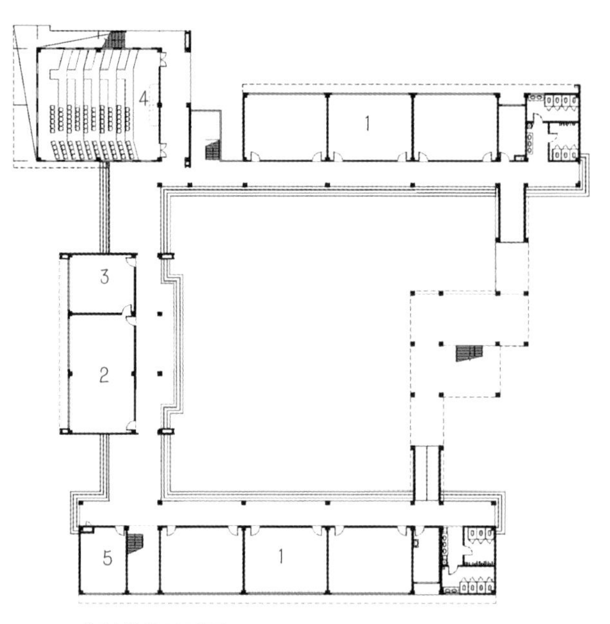

教学楼首层平面图 Ground floor plan of classroom building

1 普通教室 typical classroom
2 自然教室 science classroom
3 仪器标本准备室 instrumcnt & sample preparation room
4 225座阶梯教室 225-seat lecture hall
5 教学办公室 teaching office

建筑面向开放空间的界面采用柱廊、底层架空、屋顶平台等，形成内外空间的缓冲和联系

Colonnade, opened ground level and rooftop platform are applied in the interface of the building that faces the open space for buffering and connecting the indoor and outdoor spaces.

北川中学 Beichuan High School

援　建：中国侨联

设　计：北京市建筑设计研究院

施　工：中铁二局集团有限公司

用地面积：119 831m²

建筑面积：71 200m²

建筑规模：100班　5200人

Donator:
All-China Federation of Returned Overseas Chinese
Designer:
Beijing Institute of Architectural Design
Contractor:
China Railway No.2 Engineering Group Co., Ltd.

Site area: 119,831m²
Building area: 71,200m²
Capacity: 100 classes with 5,200 students in total

北川中学是新北川占地面积最大的公共建筑项目，位于新县城的东北角，永昌河支流云盘河从中流过。作为规模庞大的完全寄宿学校，其教学管理模式和学习生活模式的相互契合是设计方案的主要出发点。教学区、生活区和体育活动区动静分隔，疏密相间。建筑整体风格平实朴素、明朗向上，不同高低体块和立体台阶的组合表达了羌族台地建筑的意象，以羌族服饰常用的五种色彩穿插其中，地方材料的使用体现"粗粮细做"的原则。宿舍的设计强调公共交流空间的设置，阳台两两相连，并集中设置公共浴室和开水房。注重室外空间设计，为师生提供不同层次和形式的交往空间。为尊重残疾学生的心理需求，将其安排与普通学生一起生活。

Located on the northeast and enclosed a part of landscaped green belt along Yunpan River, the site of new Beichuan High School is the largest individual public service site of the new county seat. For this large-size full boarding school, its main aim of the design is to match its teaching, management, studying and living modes of the same type. It is divided into teaching area, dormitory area and sports area following the activity characteristics. The overall building style is concise and vivid; the combination of masses of different heights and three-dimensional steps express the image of terrace building of the Qiang nationality; five colors commonly used in the Qiang costume are used; the selection of local materials follows the principle of "fine design based on general materials". The design of the students' dormitory focuses on the arrangement of public communication space, with each two balconies being connected, and public shower rooms and boiler rooms are placed centrally. To respect their psychological needs, the disabled students live together with other normal students.

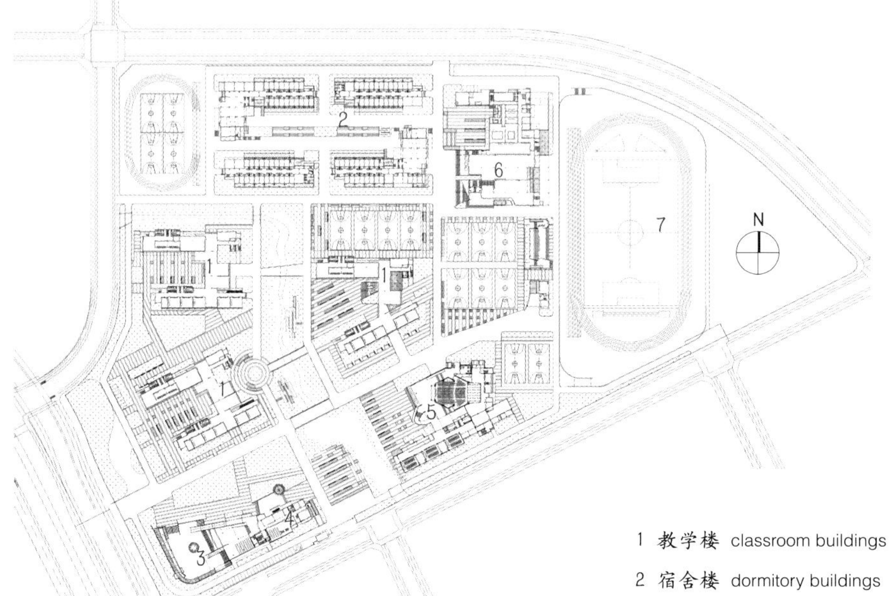

总平面图 Site plan

1 教学楼 classroom buildings
2 宿舍楼 dormitory buildings
3 纪念小广场 small memorial square
4 教学办公楼、图书馆 office building and library
5 礼堂–实验楼 auditorium and laboratory building
6 食堂 dining hall
7 运动场 sports ground

对捐建者爱心奉献的颂扬和铭记，是校园设计的重点主题之一。纪念小广场将捐赠者名单刻于纪念墙上

One of the design themes is to appreciate and memorize the donators, so a memorial wall is built on the small memorial square with inscription of the donators' names.

图书馆与教学办公楼设于同一建筑中，逐层后退的台阶，为门前的小广场提供了纪念性和观看的界面，将二者自然地联系在一起

The library and teaching offices are arranged in the same building with gradual backward steps to create memorial atmosphere and assemble spaces for the small square, and naturally connects both of them.

教学区

Teaching area

宿舍区
Dormitory area

礼堂—实验楼位于校园东南角，相对独立　The auditorium and laboratory building on the southeastern corner of the campus are relatively independent.

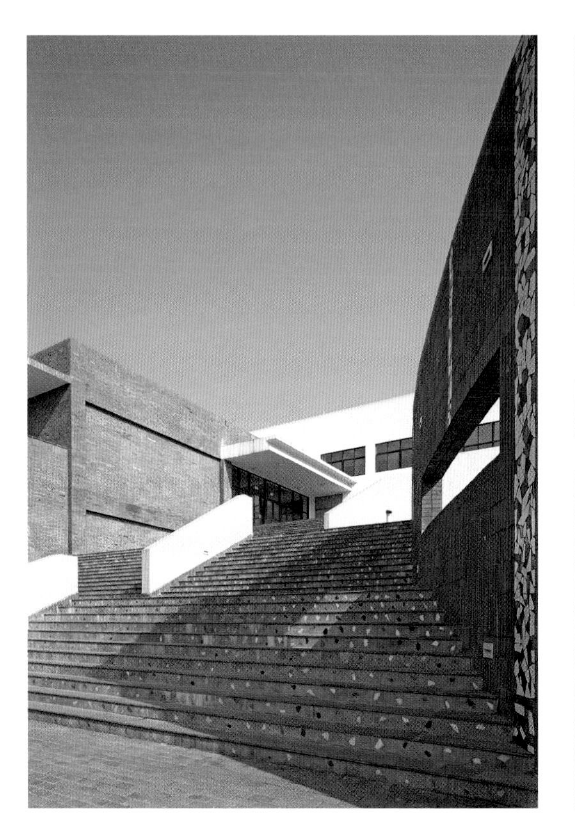

食堂位于教学区、宿舍区、操场三者之间　The dining hall is embraced by the teaching area, dormitory area and sports area.

七一高级职业中学 Qiyi Vocational High School

捐　助：中组部特殊党费

承　建：山东省聊城市

设　计：北京市建筑设计研究院

施　工：山东聊建集团总公司

用地面积：78 740m²

建筑面积：43 413m²

建筑规模：89班　4000人

Donator:

Special Party Membership Dues of CCP

Aided by:

Liaocheng, Shandong Province

Designer:

Beijing Institute of Architectural Design

Contractor:

Shandong Liaocheng Construction Group Co., Ltd.

Site area: 78,740m²

Building area: 43,413m²

Capacity: 89 classes with 4,000 students in total

七一高级职业中学位于新县城东部，南邻山东工业园区，背靠山体，由教学办公区、实训区、生活区和运动区四部分组成。教学办公区位于东侧主入口，北面实训区的机电楼、汽修楼、电子缝纫楼、旅游烹饪楼顺序南北向布置，形成垂直街道的街景空间。宿舍生活部分布置于校园西侧，内部形成台地式的丰富景观。开敞的室外运动空间则在山体和教学区之间形成减灾缓冲地带。建筑群体形成疏密有致、开阔曲折的布局，专业教学楼围合出各自的独立庭院。毛石墙面和灰瓦的运用形成简洁、现代而具有当地特色的建筑形象。

Lying in the eastern part of the new county seat, the Qiyi Vocational High School is adjacent to Shandong Industrial Park in the south and back against mountains. It consists of four areas, namely, the teaching & office area, training area, living area and sports area: the teaching & office area lies at the main entrance of the east side of the schoolyard; the training area is on the north side of the schoolyard, with electromechanical training building, vehicle repair building, electronic & sewing building and travel & cooking building aligning along the south-north direction; the living area is on the west side of the schoolyard, with its internal part forming a stepped landscape; the open outdoor sports area lies between the mountains and the teaching area, which forms a buffer zone between them. In this way, the building complex forms an open and zigzag layout with even density, each training teaching buildings enclose their independent courtyard, and the application of rough walls and grey tiles highlight concise and modern architectural image with local features.

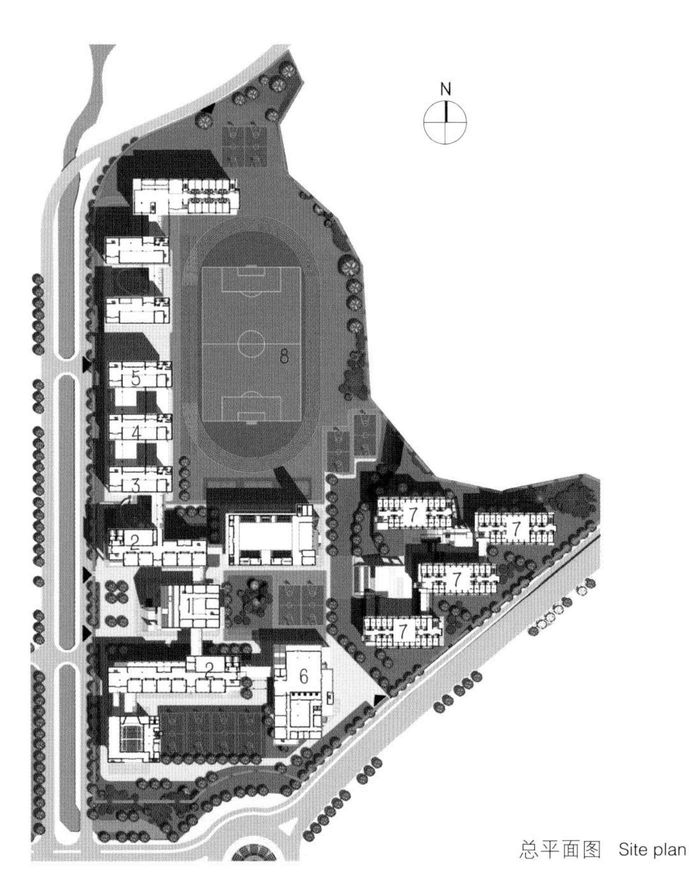

1 办公楼 office building

2 教学楼 classroom building

3 电工缝纫实训楼 electrical & sewing training building

4 机电实训楼 electromechanical training building

5 汽修实训楼 vehicle repair training building

6 食堂后勤楼 dining hall

7 宿舍楼 dormitory building

8 运动场 sports ground

总平面图 Site plan

人民医院 People's Hospital

援　建：山东省潍坊市

方　案：中国建筑标准设计研究院

施工图：潍坊市建筑设计研究院有限责任公司

施　工：潍坊昌大建设集团有限公司

用地面积：26 500m²

建筑面积：23 978m²

建筑规模：200床

Aided by:
Weifang, Shandong Province
Scheme design:
China Institute of Building Standard Design & Research
Construction drawings:
Weifang Institute of Architectural Design Co., Ltd
Contractor:
Weifang Changda Construction Group Co., Ltd

Site area: 26, 500m²
Building area: 23, 978m²
Capacity: 200 Beds

北川人民医院位于新县城西北部，尔玛小区中部。建筑采用鱼骨式布局，以两条宽度不一的医疗街串联呈枝状分布的门诊、医技、病房等功能单元，并与邻近的康复中心在二、三层相连。其中门诊、急诊、医技位于用地西侧，沿街展开，两栋病房楼位于用地东侧，南北向布置。院落空间也沿医疗街中轴线依次展开。外观设计提炼羌族传统建筑元素，通过灰砖、坡顶、碉楼、木构架、玻璃的巧妙运用，赋予建筑宜人的尺度和地方文脉的形象，避免传统医院洁净有余而亲切不足的问题。

The People's Hospital is located on the northwest of the new county seat and the middle site of Er'ma Residential Quarter. It adopts a fishbone layout where two medical streets of different widths are linked with the functional areas such as clinics, medical technology area and wards which are distributed in a branch shape, and they also connect the first and second floors of the nearby rehabilitation center. Clinics, emergency and medical technology areas are located on the west of the project site along the street; two ward buildings are arranged in a south-north direction on the east of the project site. Yard space stretches along the central axis of the medical street, and each yard is endowed with different themes featuring local garden. By skillfully utilizing grey bricks, pitched roof, turret, wood truss and glass with the traditional Qiang architectural features, the building is provided with pleasant scale and image of local cultural context.

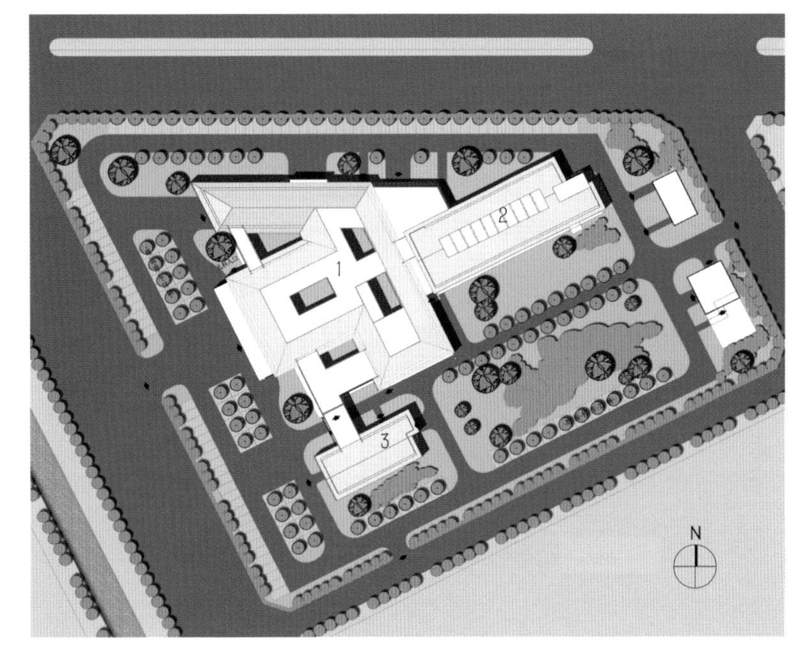

1 门诊医技楼 outpatient and medical technology building
2 住院楼 inpatient building
3 办公楼 office building

总平面图 Site plan

坡顶、灰瓦、木构架的运用改变了医院建筑通常的形象　Skillfully utilizing of grey bricks, pitched roof, turret and wood truss changes the characteristics of normal hospital buildings.

门诊采用“垂直交通——交通厅——门诊单元”的就诊流程，医疗街内设扶梯，直达各层候诊大厅

The clinic adopts a flow of “vertical traffic - traffic hall - clinic unit”. Escalators leading to the waiting halls on each floor are set in the medical street.

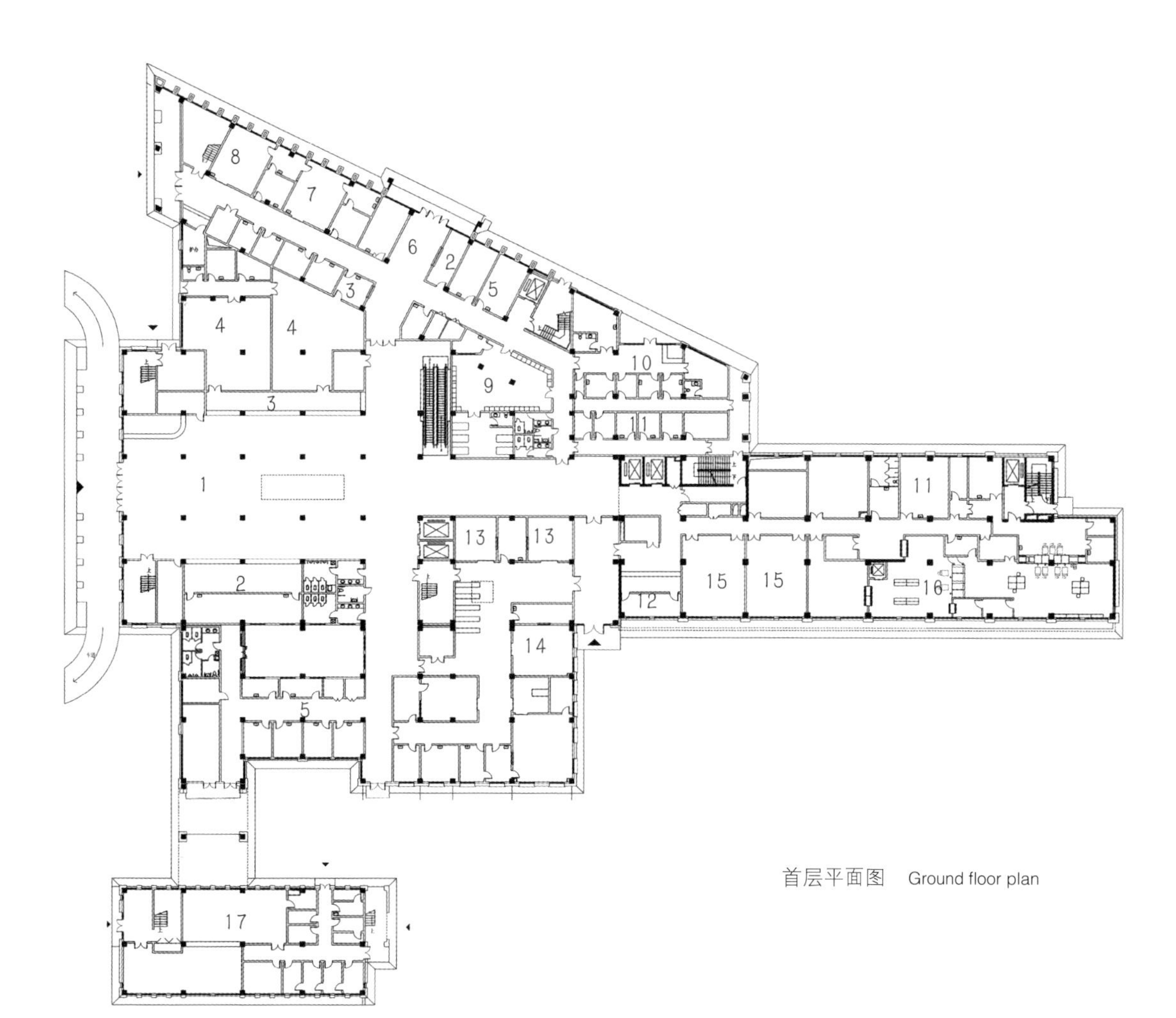

首层平面图 Ground floor plan

1 门诊大厅 outpatient hall
2 挂号、收费 registration & cashier
3 取药 medicine receiving
4 药房 pharmacy
5 诊室 consulting room
6 急诊大厅 emergency hall
7 治疗室 treatment room
8 抢救室 emergency room
9 输液室 infusion room
10 感染科 department of infectious diseases
11 值班、办公 duty room and office
12 出入院办理处 admission and discharge
13 X光室 X-ray room
14 CT室 CT room
15 药库 drug storage
16 无菌存放 sterile storage room
17 餐厅 dining hall

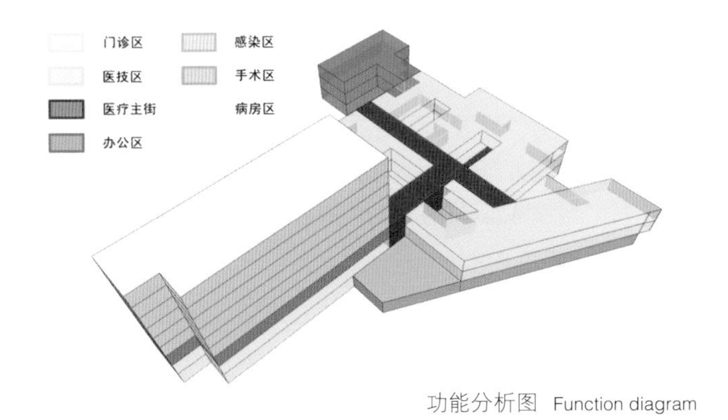

功能分析图 Function diagram

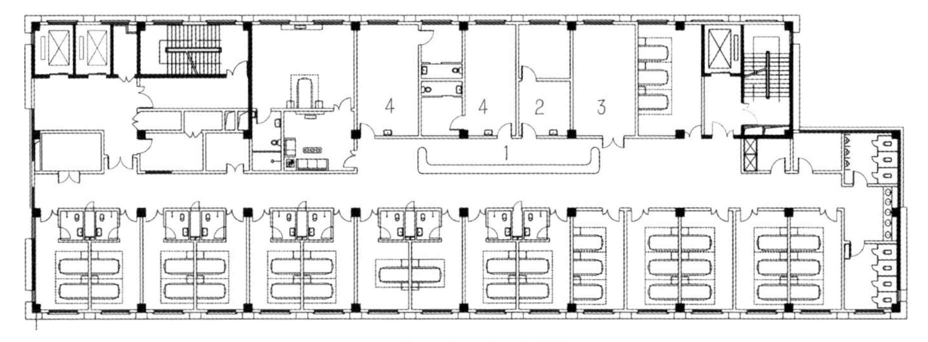

住院楼标准层平面图 Typical plan of inpatient building

1 护士站 nurse station
2 治疗室 treatment room
3 抢救室 salvage room
4 值班室 duty room

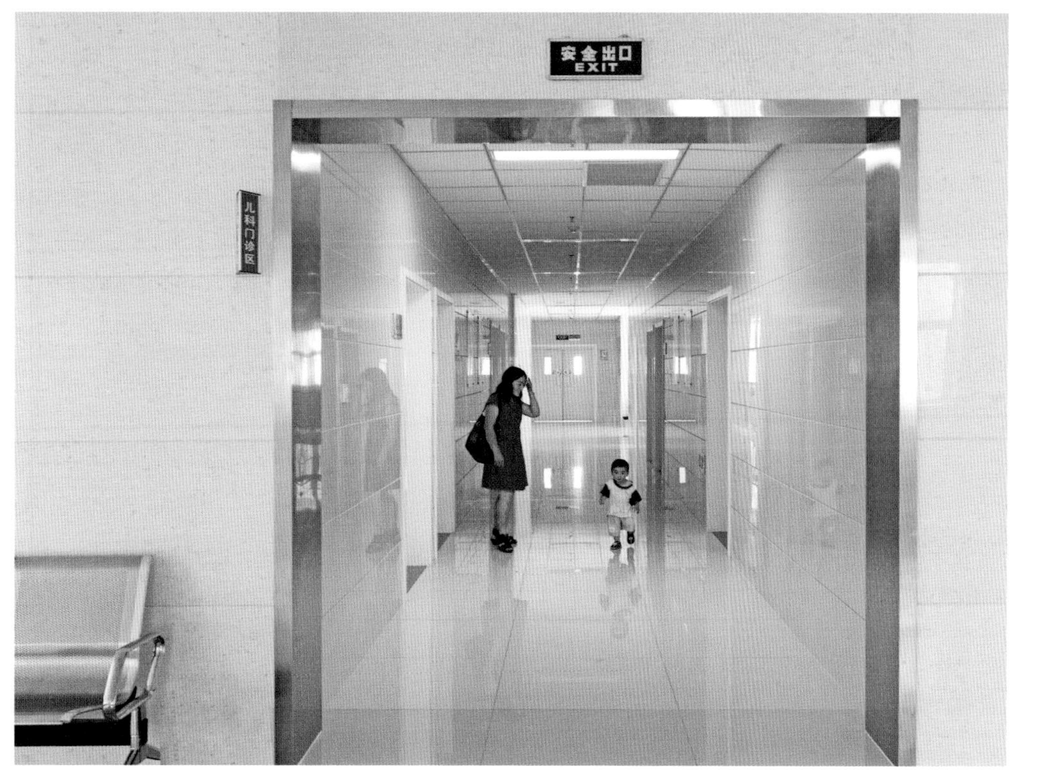
安全出口
EXIT
儿科门诊区

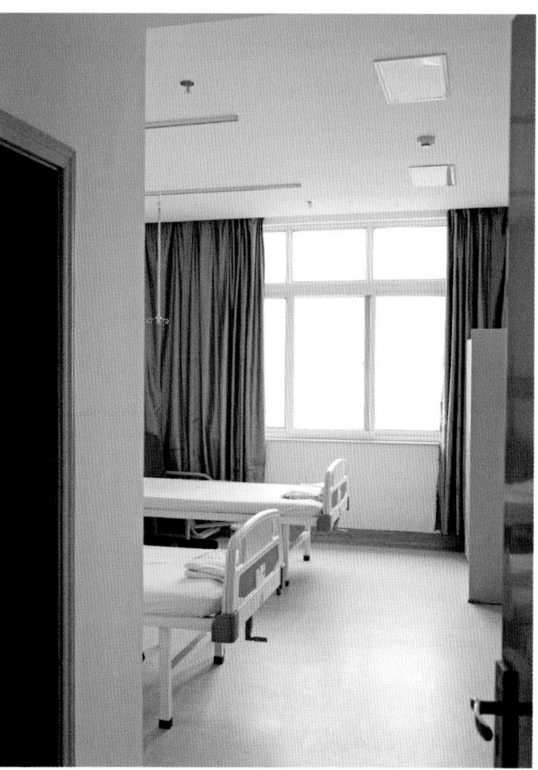

中医院、疾病预防控制中心 Chinese Medicine Hospital and Disease Prevention & Control Center

援　建：山东省菏泽市

设　计：菏泽市规划建筑设计院

施　工：山东菏建建筑集团有限公司

中医院

用地面积：8 100m^2

建筑面积：5 879m^2

建筑规模：60床

疾病预防控制中心

用地面积：5 000m^2

建筑面积：3 813m^2

建筑规模：29床

Aided by:
Heze, Shandong Province
Designer:
Heze Architectural Planning and Design Institute
Contractor:
Shangdong Hejian Construction Group Co., Ltd.

Chinese Medicine Hospital
Site area: 8,100m^2
Building area: 5,879 m^2
Capacity: 60 Beds

Disease Control Center
Site area: 5,000m^2
Building area: 3,813m^2
Capacity: 29 Beds

中医院和疾病预防控制中心处于同一地块，位于新县城西南部，新川小区北侧。中医院设于地块南部，沿道路横向展开。疾控中心根据功能要求分为几座单体建筑，综合用房与保障用房用走廊联系起来，与检验用房分开，并围合形成面向东侧的入口广场。立面处理以汉羌建筑为原型，屋顶为多重组合的坡屋顶，采用垂花门、雕花窗等细部语汇修饰，并突出片石的肌理和简洁的白色墙面的对比。

Chinese Medicine Hospital and Disease Prevention & Control Center are located in the same site on the south of the new county seat, which is close to the Xinchuan Residential Quarter. The Chinese Medicine Hospital is developed along the main road on the south side, while the Disease Control Center encloses its own entrance square facing east with its complex building and the inspection building. Multiple combinations of pitched roofs are adopted, and the traditional elements such as festoon gate and carved windows in details and the contrast of rough texture and white wall are used to emphasize the characteristics of Han-Qiang style.

北川中

妇幼保健院 Maternal & Child Health Hospital

援　建：山东省菏泽市

设　计：菏泽市规划建筑设计院

施　工：山东菏建建筑集团有限公司

用地面积：4 500m²

建筑面积：3 500m²

建筑规模：36床

Aided by:
Heze, Shandong Province
Designer:
Heze architectural Planning and Design Institute
Contractor:
Shangdong Hejian Construction Group Co., Ltd.

Site area: 4,500m²
Building area: 3,500m²
Capacity: 36 Beds

妇幼保健院位于新县城东部，永昌小区内。由于地块呈三角形，建筑布局结合地形转折，自然形成主体建筑面南，辅楼位于西侧的围合布局。与中医院及疾病控制中心的风格一致，妇幼保健院的空间组合和形式同样延续汉羌风貌的原型，以片石肌理、白色涂料、实木格栅和青灰色的屋顶体现简洁并具有地域性的建筑特征。

The Maternal & Child Health Hospital is situated on the east side of the new county seat within the Yongchang Residential Quarter. The layout, in which the main building faces the south side, and the auxiliary building lies on the west side, suits the triangle site. With the same style as Chinese Medicine Hospital and Disease Control Center, it also follows the prototype of local traditional housing features, such as rough texture, white coating, solid wood grille and grey roofs.

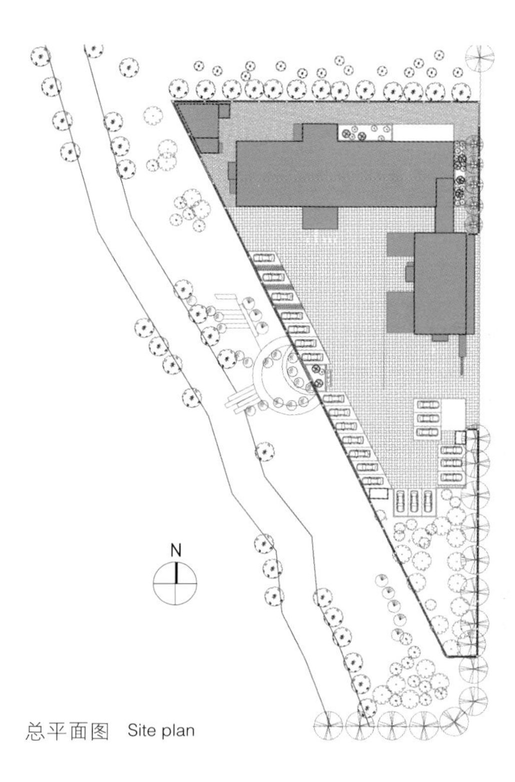

总平面图 Site plan

老年活动中心 Senior Citizen Activity Center

援　建：山东省枣庄市

设　计：枣庄市建筑设计研究院

施　工：滕州市建筑安装工程集团公司

用地面积：4 900m^2

建筑面积：3 442m^2

Aided by:
Zaozhuang, Shandong Province
Designer:
Zaozhuang Institute of Architectural Design and Research
Contractor:
Tengzhou Construction and Installation Engineering Group Corporation

Site area: 4,900m^2
Building area: 3,442m^2

老年活动中心位于新县城北部，东邻永昌河景观带。平面布局强调动静分区，分为三部分，北侧安排图书、棋牌、美术等安静的活动，南为运动、音乐、曲艺活动，主入口部分衔接二者，并设置办公、接待空间。建筑的主要出入口均设无障碍坡道，采用外廊将各个房间串联起来，为视力有障碍的老年人提供明亮通畅的交通空间。每层均设残疾人专用卫生间。建筑风貌体现现代羌风，以坡屋顶为主，高低错落，并采用片石外墙和仿木构件，整个建筑形态舒缓，色彩柔和，适应老年人的感受需求。

The Senior Citizen Activity Center is located on the north side of the new county seat, and to the east of the landscaped green belt along Yongchang River. Its layout focuses on a division of static and dynamic area, and includes three parts: the north part is for quiet activities such as reading, playing chess and painting, the south part is for sports, music and folk arts, and the main entrance is connected with the former two with offices and reception space. Besides, the entrances and exits of the center are set with accessible ramps, leading to verandas which link each room and provide bright space for the visually impaired aged people. Each floor is set with a special toilet for the disabled. The overall building style reflects the modern Qiang features with pitched roofs of different heights. By applying external rough walls and wood-like components, the whole building appears cozy in gentle colors to meet psychological need of the senior citizens.

1 门厅 lobby
2 小卖部 shop
3 咨询室 counseling room
4 曲艺教室 folk arts classroom
5 音乐教室 music classroom
6 乒乓球室 table tennis room
7 台球室 billiards room
8 图书室 reading room
9 书库 book storeroom

首层平面图 Ground floor plan

40

社会福利中心 Social Welfare Center

援　建：山东省临沂市

方　案：中国汉嘉设计集团有限公司

施工图：临沂市建筑设计研究院

施　工：天元建设集团有限公司

用地面积：15 749m²

建筑面积：7 226m²

建筑规模：福利院 200床

养老院 100床

未成年人保护中心及救助站 20床

Aided by:
Linyi, Shandong Province
Scheme design:
China Hanjia Desgin Group Co., Ltd.
Construction drawings:
Planning & Architecture Design Institute of Linyi
Contractor:
Tianyuan Construction Group Co., Ltd.

Site area: 15,749m²
Building area: 7,226m²
Capacity:
Welfare House 200 Beds
Senior Citizens' Home 100 Beds
Children & Adolescents Protecting Center and First Aid Station 20 Beds

社会福利中心由福利院、敬老院、未成年保护中心、救助站四部分组成，同处于新县城北部的一个狭长地块。基地四面环山，建筑群分为南北两组。福利院及公共服务部分位于地块北侧，通过连廊把各种功能串联起来，形成内向型庭院和均匀分布的组团景观，便于使用者的通达和交流。南部建筑亦分为敬老院、未成年人保护中心及救助站两组，食堂等服务设施分设，避免老年人与未成年人之间的干扰。建筑外观简洁、明快，以材料和色彩体现羌族风格。

The Social Welfare Center is composed of the Welfare House, the Senior Citizens Home, the Children & Adolescents Protecting Center and the First Aid Station. The project is located on a long-narrow site north to the new county seat, where is surrounded by mountains, with its building complex being divided into two groups. Of which, the Welfare House and service facilities are located on the north side, with their functions connecting through corridors to form inward-oriented courtyard and evenly distributed group landscape, which are convenient for users to access and communicate; the buildings on the south side are divided into the group of the Senior Citizens' Home, and the group of the Children & Adolescents Protecting Center and the First Aid Station, while their dining rooms are respectively set to avoid disturbances between the elderly and under-aged people. The appearance of buildings is concise and delightful, which reflects Qiang style by means of material and color.

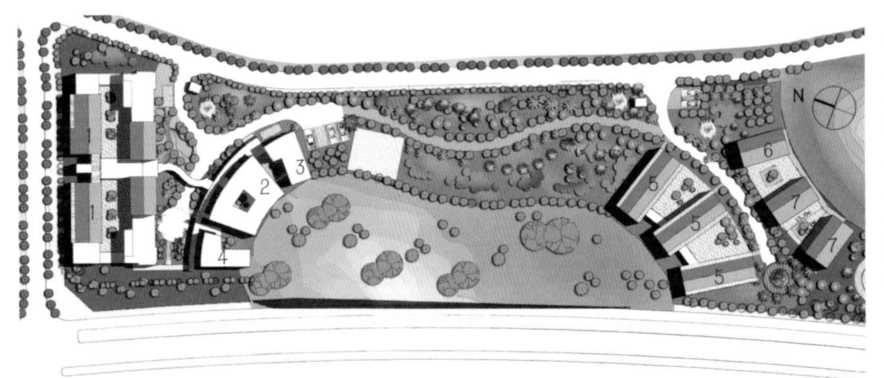

1 福利院 Welfare House
2 食堂 dining hall
3 诊所 clinic
4 浴室 bath house
5 养老院 Senior Citizens' Home
6 救助站 First Aid Station
7 未成年人保护中心 Children & Adolescents Protecting Center

总平面图 Site plan

工人俱乐部 Workers' Club

方　案：清华大学建筑设计研究院

施工图：四川海辰工程设计研究有限公司

施　工：四川华西集团有限公司

用地面积：2 674m²

建筑面积：3 405m²

Scheme design:
Architectural Design and Research Institute of Tsinghua University
Construction drawings:
Sichuan Haichen Engineering Design Co., Ltd.
Contractor:
Sichuan Huaxi Group Corporation Limited

Site area: 2,674m²
Building area: 3,405m²

工人俱乐部紧邻开发区服务中心，用于满足开发区工人文娱活动和专业培训的需求。平面布局为L形，沿道路转角设置，以单廊形式组织功能空间，在其后形成内院布置篮球场和停车位。外观设计遵循工业园区现代、简洁的需求，采用与周边开发区服务中心立面相协调的玻璃、涂料和暖色面砖的搭配，营造出明快、和谐的效果。窗外设置的木质百叶则为建筑增添了地方元素。

Close to the Service Center of Beichuan Economic Development Area, the Workers' Club is built for cultural events, entertainment and professional training of workers in the development area. The plan layout of the club adopts an L shape along the road corner; with a single corridor ahead, the club has a basketball court and a parking lot in the inner courtyard behind the building. Its appearance design follows the modern and simple demands of the industrial park, adopting a collocation of glass, coating and warm-color face tiles that are in line with the façade of the Service Center, to create bright and harmonious effect. Besides, the wooden window blinds reflect local elements for the buildings.

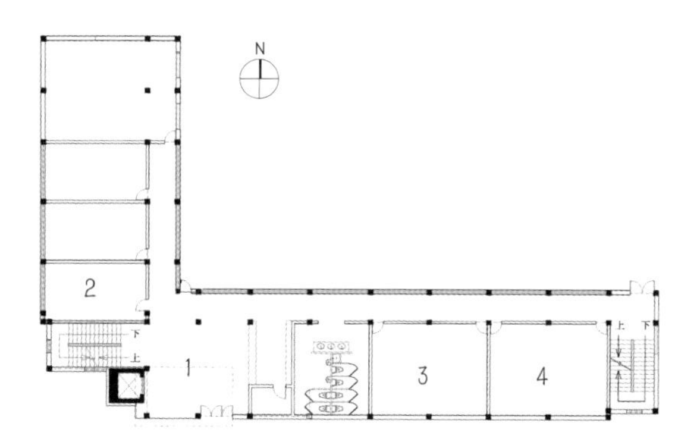

1 门厅 lobby
2 办公室 office
3 困难职工帮扶中心 workers aid center
4 职工职业服务中心 workers employment service center

首层平面图 Ground floor plan

文化轴线 精神家园

Cultural Axis and Spiritual Home

北川新县城作为国内唯一羌族自治县——北川县的大灾之后异地重建的全新城镇，新县城文化轴线的规划设计饱含了弘扬民族文化，重建精神家园的文化复兴期冀。

北川新县城新址周边群山环绕，四水中流，山水环境特质突出。文化轴线的走向布局正是出于融合新县城周边自然山势与河道水形的考虑，通过借鉴羌族传统村寨依山而生，近水而居的营造构思，强调巧于因借，通达山水的设计理念。

而在环、廊、带、脊、轴、链相互交织的北川新县城空间结构中，文化轴线也是唯一与其他结构要素均有交织的核心空间要素。它以云盘山为发端，由东向西延伸，跨越安昌河与安北路，遥指远处群山。

轴线主体串联布局北川文化中心（含博物馆、文化馆与图书馆三馆）、抗震纪念园（含静思、英雄、幸福三园）、羌族特色步行街（含禹王广场）以及禹王桥等功能节点，两侧富集新川路商业街、北川影剧院、新北川宾馆、永昌镇政府等县级服务设施。规划设计力求以山居、营寨、追思、祈福、重建、祝福、繁荣的清晰主题序列，融汇时间标尺与文化传承，倾注民族情感与精神寄托，将文化轴线区域打造为城建工程标志的核心、抗震精神标志的强音以及文化遗产标志的亮点。

选择轴线这样一种秩序清晰、感知强烈的空间形式也是规划设计兼顾平衡特色与重建效率考量，期冀形成空间秩序特色的同时，为快速重建背景下的城镇发展提供有序、高效的城镇空间格局。（李明 执笔）

As a relocation of Beichuan, the only Qiang autonomous county in China, the new county seat is designed with a cultural axis which bears anticipation of carrying on the ethnic culture and rebuilding the spiritual home.

Surrounded by mountains with rivers running through, the new county seat of Beichuan has distinctive environmental characteristics. The layout of the cultural axis is determined by considering the natural mountain terrain and shape of the river course around the new county and by using the idea of building a traditional Qiang village against the mountains and close to water, stressing the design concept of skilfully utilizing the local conditions for well-organized landscape design.

In the spatial structure of the new county seat of Beichuan featured by an intertwinement of circle, corridor, belt, ridge, axis and chain, the cultural axis is the only core spatial factor having connection with other structural elements. Starting from the Yunpan Mountain, the cultural axis stretches from the east to the west, and runs cross the Anchang River and Anbei Road, facing the mountains afar.

The main body of the cultural axis connects such functional nodes as the Cultural Center (including the Qiang Folk Museum, Cultural Hall and Library), the Memorial Park (the Meditation Garden, the Hero Garden and the Happiness Pavilion), the Qiang Features Pedestrian Street (Yuwang Plaza included) and Yuwang Bridge, with county-level service facilities including the Xinchuan Road Commercial Street, Arts Center, New Beichuan Hotel, Government of Yongchang Town arranged on both sides of the axis. The planning and design are themed with neighboring mountains, village, recollection, praying, reconstruction, blessing and prosperity. By blending time scale and cultural heritage, and with national emotion and spiritual support, it aims to build a cultural axis area into "a core symbol of urban construction, a symbol of quake relief spirit and a landmark of cultural heritage symbol".

Axis is a spatial form that has clear order and is perceptive, and it is selected in the design after considering the balance feature and reconstruction efficiency, hoping to form a distinctive spatial order, and providing orderly and efficient town spatial layout for the town development against the background of rapid post-quake reconstruction. (by Li Ming)

2010年建设中的文化轴线（中规院供稿）

The Cultural Axis under construction in 2010 (by CAUPD)

禹王桥
羌族特色步行街
禹王广场
抗震纪念园
北川文化中心

禹王桥 Yuwang Bridge

援　建：山东省青岛市

方　案：成都富政建筑设计有限公司

施工图：青岛市市政工程设计研究院

施　工：青岛第一市政工程有限公司

建筑面积：5 166m²

长　　度：204.2m

宽　　度：12.6m

Aided by:
Qingdao, Shandong Province
Scheme design:
Chengdu Fuzheng Architectural Design Co., Ltd
Construction drawings:
Qingdao Municipal Engineering Design & Research Institute Co., Ltd.
Contractor:
Qingdao No. 1 Municipal Construction Engineering Co., Ltd.

Building Area: 5,166m²
Length: 204.2m
Width: 12.6m

禹王桥是一座人行风雨廊桥，位于新县城文化轴线上，横跨安昌河，是新县城景观轴和步行廊道的西侧起点，集中体现了传统羌族特色风貌。立面设计充分表达羌族风情，三段式划分，避免了桥体过长的连续面带来的单调感。禹王桥内设置了商业店铺空间，并为行人提供了驻留、休憩、观景的平台。坡屋顶间设置的高窗，利用烟囱效应为相对封闭的廊桥提供了良好的自然通风环境。

The Yuwang Bridge is a covered bridge which lies on the cultural axis of the new county seat and goes across the Anchang River. As one significant part of the landscape axis and pedestrian walkway, the bridge showcases the style and features of the Qiang ethnic group. The elevation is divided into three parts to avoid dullness produced by continuous surface of the long bridge. Inside the bridge, there are also shops, benches and sight view platforms. The clearstories which are set in the pitched roof can provide good and natural ventilation for the relatively enclosed bridge in a stack effect.

禹王桥在造型上融入传统羌族的索桥、笮桥和碉楼形象　The Yuwang Bridge incorporates the traditional Qiang elements such as cable bridge, bamboo suspension bridge and turret.

碉楼的厚重体量强化了新县城的"门户"空间

The massive turret enhances the "gateway" space of the new county seat.

羌族特色步行街 Qiang Features Pedestrian Street

援　建：山东省济宁市、威海市

方　案：北京清华城市规划设计研究院

成都富政建筑设计有限公司

青岛市建筑设计研究院股份有限公司

施工图：济宁市建筑设计研究院

济南中建建筑设计院有限公司

施　工：山东圣大建设集团有限公司

山东宁建建设集团有限公司

山东省建设集团有限公司

用地面积：75 600m²

建筑面积：70 000m²

Aided by:
Jining and Weihai, Shandong Province
Scheme design:
Urban Planning & Design Institute of Tsinghua University
Chengdu Fuzheng Architectural Design Co., Ltd
Qingdao Architectural Design & Research Institute Co., Ltd
Construction drawings:
Jining Architecture Design & Research Institute
CSCEC Jinan Architectural Design Institute Co., Ltd.
Contractors:
Shandong Shengda Construction Group Co., Ltd.
Shandong Ningjian Construction Group Co., Ltd.
Shandong Provincial Construction Group Co., Ltd.

Site area: 75,600m²
Building area: 70,000m²

羌族特色商业街是新县城景观轴和步行廊道的重要组成部分，是集中体现传统羌族风貌特色的重点区域。旅游业将成为新北川的支柱产业之一，步行商业街因此承担了创造多样化、多层次的旅游体验的任务。

由于原生态传统羌式建筑的内部空间已不能满足现代城市生活所需，因此商业街的建筑定位为“仿原生传统羌式建筑”。外部形象充分体现原生态羌族建筑特色，建筑体量化整为碎，均为2～4层，造型多变，屋顶平坡结合，通过碉楼、廊桥和小广场的设置，形成丰富的天际线，并采用片石、块石、原木等传统建筑材料，力求原汁原味。建筑内部空间则按商业、餐饮、休闲娱乐、旅游接待等现代功能设计，结构上也采用现代结构体系。

Qiang Features Pedestrian Street is one significant part of the landscape axis and pedestrian walkway of the new county seat, which demonstrates the traditional Qiang style and features. The design focuses on developing “experimental tourism” in an attempt to make tourism be one of the mainstay industries and build human-concerned public space. The general layout of the shopping street is designed as a business-orientated mode which can produce employment opportunities of multiple types and levels.

Since the interior space of the original Qiang buildings no longer caters to demands of modern urban life, the buildings on the shopping street are defined as “simulation of the original Qiang architecture”, which demonstrate the original flavor of the traditional Qiang houses from the appearance while endow its interior space with modern functions such as commercial, food & beverage, recreation, entertainment and tourism and also with a design of modern structural system.

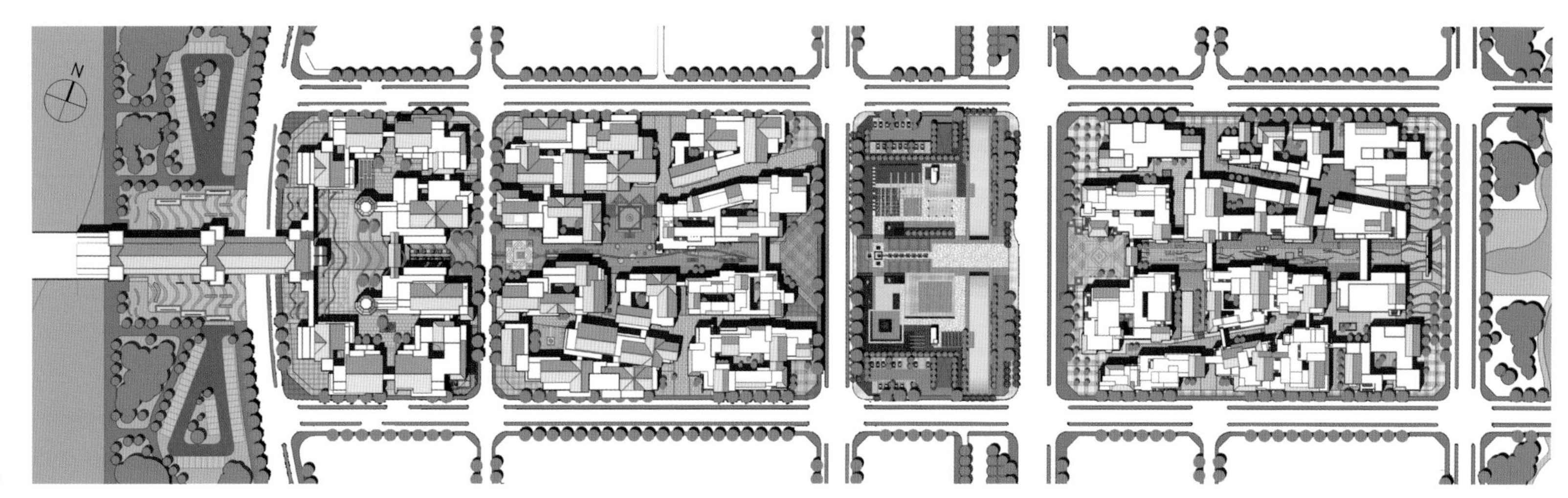

总平面图　Site plan

北川羌族特色步行街
宁圣大建设集团
济宁建设监理有

India
天竺五坊
老家味面

步行街西段的碉楼，作为禹王广场的视觉控制点，也丰富了整个建筑群的天际线　The turret of west part of the street focuses the view on the Yuwang Plaza, and enriches the skyline of the whole building group.

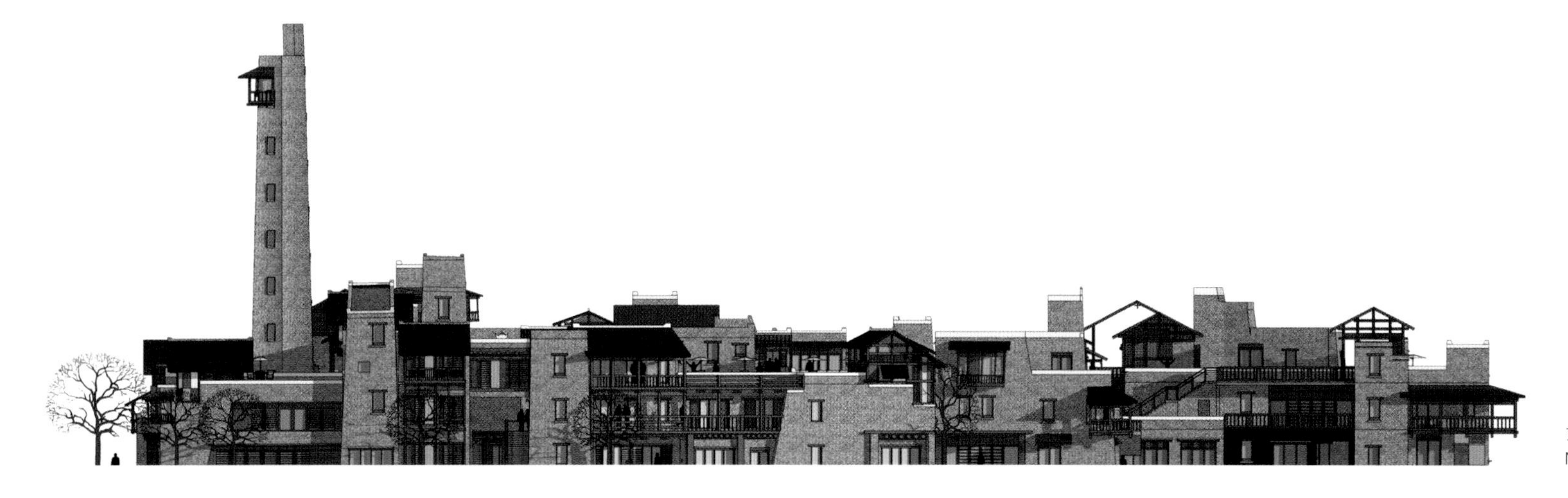

步行街西段西北立面
Northwest elevation of west part of the street

风雨廊桥、平台、连廊、室外楼梯的串联，形成立体而多样的商业空间

Covered bridges, platforms, galleries and outdoor stairs are working together to form a three-dimension commercial system with various spatial experiences.

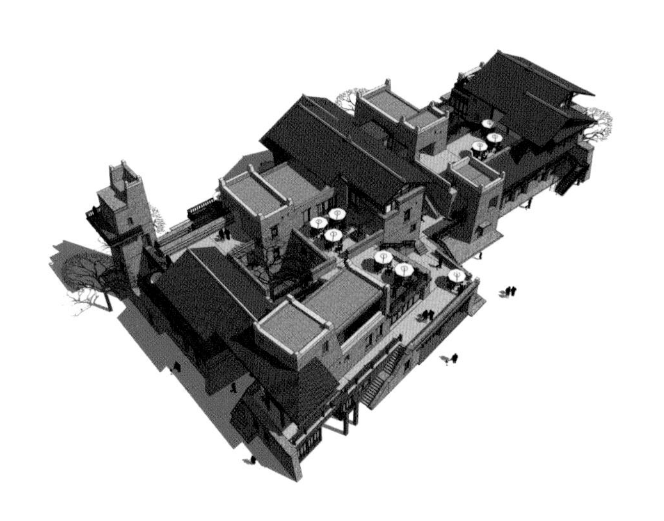

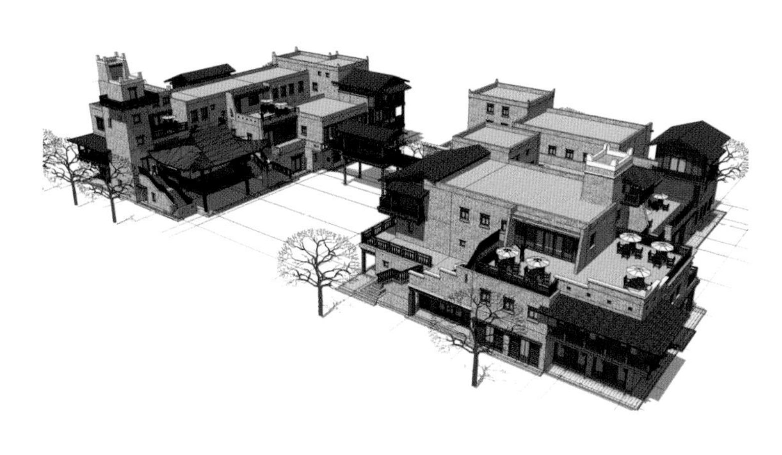

“巴拿恰”即羌语“商贸街”之意

Another name of the street – Banaqia, means “shopping street” in Qiang dialect.

步行街的东端北侧，设有北川非物质文化遗产中心，介绍羌民族丰富的传统文化

Beichuan Intangible Cultural Heritage Center, located on the northeast end of the shopping street, presents the traditional culture of the Qiang ethnic group.

抗震纪念园 Memorial Park

援　建：山东省青岛市

设　计：总　图 中国城市规划设计研究院

静思园 天津华汇工程建筑设计有限公司

幸福馆 清华大学建筑设计研究院

英雄园 深圳市建筑设计研究总院有限公司

园　林 北京北林地景园林规划设计院

照　明 北京光景照明设计有限公司

设计总指导 张锦秋

设计总协调 宋春华

施　工：莱西市建筑总公司

青岛第一市政工程有限公司

青岛市黄岛园林绿化工程有限公司

四川华西集团有限公司

Aided by:
Qingdao, Shandong Province

Designers:
Site design: China Academy of Urban Planning & Design
Meditation Garden: Tianjin Huahui Architectural Design & Engineering Co., Ltd,
Happiness Pavilion: Architectural Design and Research Institute of Tsinghua University
Hero Garden: Shenzhen General Institute of Architectural Design and Research Co., Ltd
Landscape: Beijing Beilin Landscape Architecture Institute Co., Ltd.
Lighting: Light & View Lighting Design Co., Ltd.
General design director: Zhang Jinqiu
General design coordinator: Song Chunhua

Contractors:
General Construction Company of Laixi
Qingdao No. 1 Municipal Construction Engineering Co., Ltd.
Qingdao Huangdao Landscaping & Afforesting Co., Ltd.
Sichuan Huaxi Group Corporation Limited

北川抗震纪念园是北川新县城灾后重建、抗震精神的重要标志场所，位于北川新县城文化轴线和景观轴线的交汇处。规划设计反映了地震、救灾和重建的全过程，以“静思园”、“英雄园”和“幸福园”的组合体现了“追思灾害”、“纪念抗震”和“展现幸福”的设计定位。

Lying on the central part and at the intersection of the cultural axis and landscape axis of the new county seat, the Memorial Park is the important symbolic spot to represent the spirit of earthquake relief and reconstruction. Its planning reflects the overall process of earthquake, disaster relief and reconstruction, the park is designed to “recollect earthquake, commemorate relief work and show happy life” and it includes three gardens – “Meditation Garden”, “Hero Garden” and “Happiness Garden”.

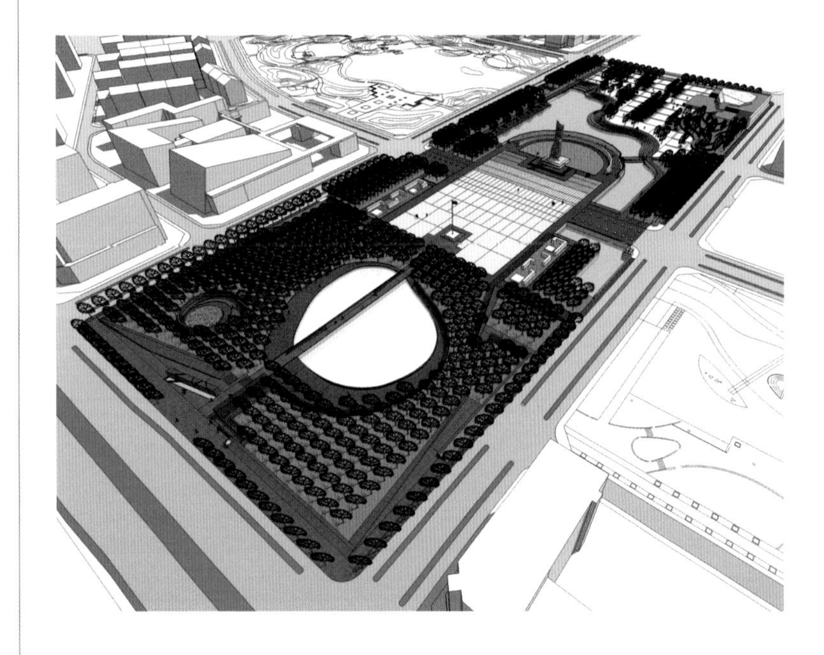

用地面积：57 000m²

Site area: 57,000m²

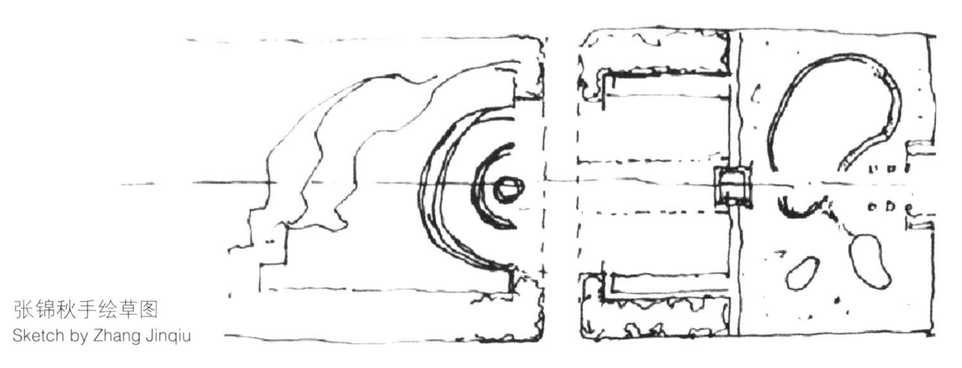

张锦秋手绘草图

Sketch by Zhang Jinqiu

英雄园主题雕塑
《新生》
叶毓山 作

Theme Sculpture of
Hero Garden *Rebirth*
(by Ye Yushan)

静思园——体现记忆与思念

“静思园”位于抗震纪念园东侧，以纪念灾难、缅怀同胞为主题，利用简洁、现代的手法营造安静、幽闭的空间氛围，为人们追思过去提供沉思的空间，其下沉空间、浮雕墙面和静止水系等空间要素体现记忆与思念的空间定位。中间的下沉步行路径，将人流限定在更加幽静的空间中以突出安静、幽闭的氛围。路径穿过一片静止的水面，水面形式如同一滴眼泪，表达对灾难和逝者的哀思。以自然砂石、灰砖、青铜为材料，5m间隔的银杏树阵，营造规整茂密的林下空间，提供漫步沉思的想象空间。

用地面积：16 200m^2

Meditation Garden — memorizing and missing

The Meditation Garden, located on the east side of the Memorial Park, is concise and quiet in its modern design method in an attempt to record the disaster and memorize the earthquake victims. The architectural elements, such as sunk walkway, embossment wall and quiet water surface form a memorizing and missing space. The pedestrian path sinks to improve a sense of enclosure and embrace pedestrians in a more tranquil place to highlight its theme. The water surface of the Meditation Garden appears as tear dropping to express deep recollection of the disaster and the deceased. In the garden, square grids of 5m for each are planted with ginkgoes where people can freely pass through in the dense woods. At the same time, natural gravels, grey bricks and bronzes are used to keep purity of the Meditation Garden.

Site area: 16,200m^2

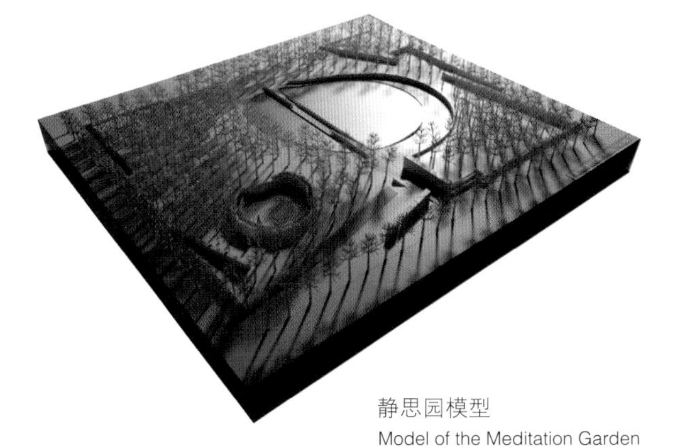

静思园模型

Model of the Meditation Garden

英雄园——展现纪念与歌颂

“英雄园”位于抗震纪念园中部，以赞颂和弘扬抗震救灾和重建精神为主题，利用对称、序列的手法营造严肃、庄重的氛围。7000m²的广场可同时容纳4000人，成为北川新县城大型纪念活动、节日集会和集体文娱活动的重要开敞空间。英雄园西侧以半圆形草坡广场与幸福园的流动水系衔接，圆心位置安排大型主题雕塑“新生”。雕塑高约25m，是整个纪念园及周边区域的视觉焦点。广场的地面铺装纹理变化由疏到密，增加了广场的进深感，进一步烘托了主题雕塑的挺拔效果。

用地面积：15 800m²

Hero Garden — commemorating and eulogizing

The Hero Garden is located in the middle of the memorial park, which is built for commemorating and eulogizing the spirit of earthquake relief and reconstruction. By using symmetrical and sequential methods to create a grand and open atmosphere, the 7000m² square can accommodate 4,000 people as a large urban space for assembling activities. The semicircular grassland square in the west side connects with the flowing water system of the Happiness Garden, and in the center erects a theme sculpture. The 25m-high sculpture "Rebirth" becomes a visual focus of the whole memorial park and surrounding areas. The paving patterns on the square floor change from sparse to dense, which can consolidate the deep sense of the square, and further set off magnificence of the sculpture.

Site Area: 15,800m²

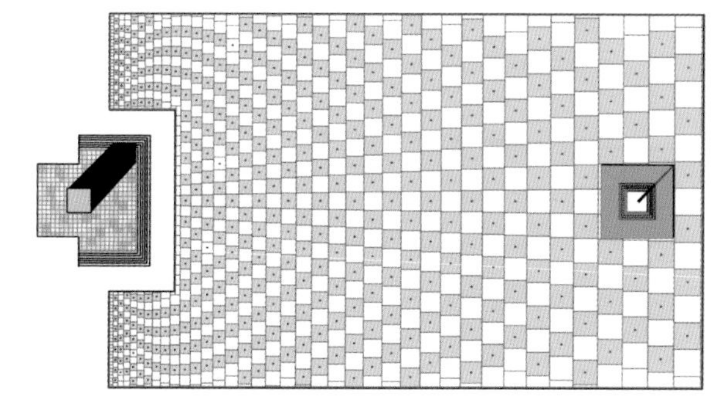

英雄园铺地设计 Paving design of the Hero Garden

幸福园——营造重生与欢乐

“幸福园”是以展现新生活、新希望为主题，利用自由、曲折的空间线条，营造活泼、亲切的空间氛围，作为群众活动休闲的重要场所，营造重生与欢乐的美好景象。开阔的水面增加了空间的流动效果，滨水地带采用亲水的处理手法，将水系景观最大限度地引入幸福园内。

用地面积：25 000m²

Happiness Garden — rehabilitating and rejoicing

Located on the west side of the memorial park, the Happiness Garden faces the Hero Garden across a waterscape. By applying free and flexuous lines in a lively and amiable atmosphere, the design aims to present a new life and new hope. In the garden, the waterfront belt forms a receding platform for sight view, which allows people to get closer to the water.

Site Area: 25,000m²

幸福馆

幸福园展览馆是幸福园中的主体建筑，集中展示了北川县社会经济文化等各方面的发展和愿景，是“再造一个新北川”的精神的延续。建筑以羌族图腾“白石”作为主题，暗喻神圣、庇护之意。大部分建筑体量置于地下和半地下，采用非对称布局，面向中心及水面方向进行适当退让，与环境融为一体。开敞的景观平台与广场绿地相结合，提供了亲切、和谐的城市公共空间。

建筑面积：2 353m²

Happiness Pavilion

As the main building of the Happiness Garden, the Pavilion exhibits the development and planning of social, economic and cultural aspects in Beichuan County. It is the physical representation of “Building a New Beichuan”. The hall is designed as the “white stone” image of the Qiang's belief in totem, conveying its sanctity and protection. Most of the building is set underground or semi-underground and adopts a non-symmetrical layout with an appropriate setback facing the center and the water surface to create a harmonious combination of the garden with the landscape. The open landscape platform, together with the square green area, forms comfortable and harmonious public space of the town.

Building area: 2,353m²

剖透视图 Sectional Prospective

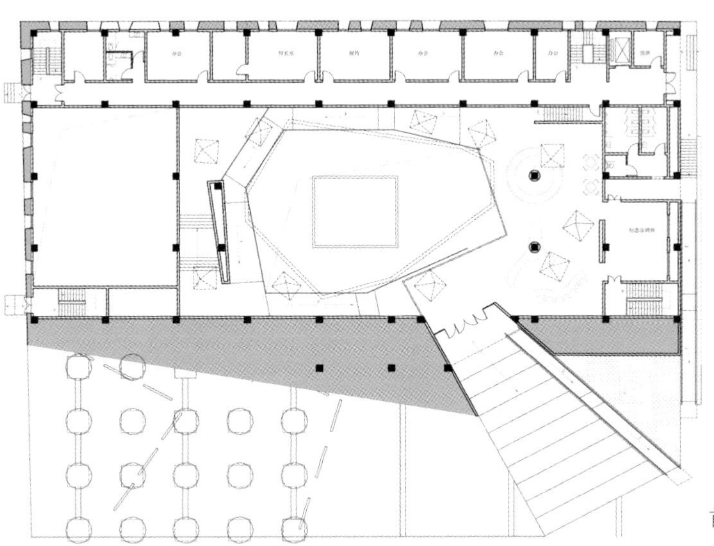

首层平面图 Ground floor plan

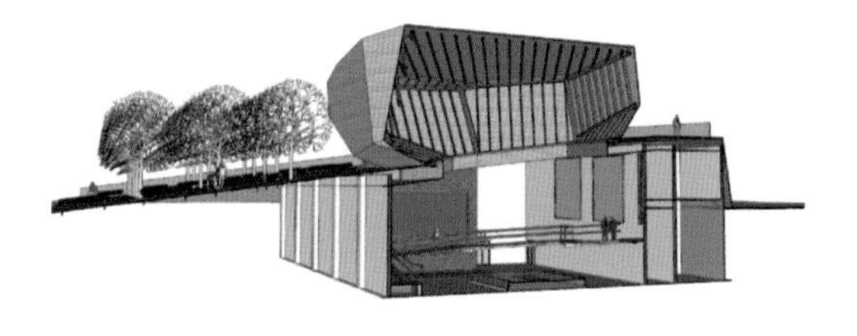

文化中心 Cultural Center

设　计：中国建筑设计研究院

施　工：山东德建集团有限公司

用地面积：22 438m²

建筑面积：14 098m²

羌族民俗博物馆

捐　助：澳门基金会

承　建：山东省德州市

建筑面积：8 002m²

图书馆、文化馆

援　建：山东省德州市

建筑面积：图书馆 3 088m²

文化馆 3 008m²

Designer:
China Architecture Design & Research Group
Contractor:
Shandong Dejian Group Co., Ltd.

Site area: 22,438m²
Building area: 14,098m²

Qiang Folk Museum
Donator: Macao Foundation
Aided by: Dezhou, Shandong Province
Building area: 8,002m²

Library and Cultural Hall
Aided by: Dezhou, Shandong Province
Building area: Library 3,088m²
Cultural Hall 3,008m²

北川文化中心位于新县城中轴线东北尽端，与抗震纪念园相邻，由图书馆、文化馆、羌族民俗博物馆三部分组成。设计构思源自羌寨聚落，以起伏的屋面强调建筑形态与山势的交融，建筑作为大地景观，自然地形成城市景观轴的有机组成部分，并与城市背景获得了巧妙的联系。开敞的前庭既连接三馆，也可作为各族人民交流聚会的城市客厅。建筑以大小、高低各异的方楼作为基本构成元素，创造出宛如游历传统羌寨般丰富的空间体验。碉楼、坡顶、木架梁等羌族传统建筑元素经过重构组合，成为建筑内外空间组织的主题，并强调了与新功能和新技术的结合。

Situated in the northeast of the central axis of the new county seat, the Cultural Center is adjacent to the Memorial Park, and composed of Library, Cultural Hall and Qiang Folk Museum. The design conception for the Cultural Center originates from the Qiang settlement, and emphasizes on the blending of the architectural form and mountains with fluctuant roofing; as the earth landscape, the building naturally forms an integrated component of the urban landscape axis, and subtly contacts with the urban background. The open forecourt not only connects the three pavilions above-mentioned, but also can be considered as a town hall in which people from all nationalities can convene and communicate together. In addition, taking square buildings of various sizes and heights as the basic element, the building complex creates an abundant space experience as if people are travelling in a Qiang village. Through reconstruction and combination, the Qiang traditional architectural elements such as turret, pitched roof and wooden beam become the theme of the building's internal and external spatial organization, which emphasize on the combination with new functions and technologies.

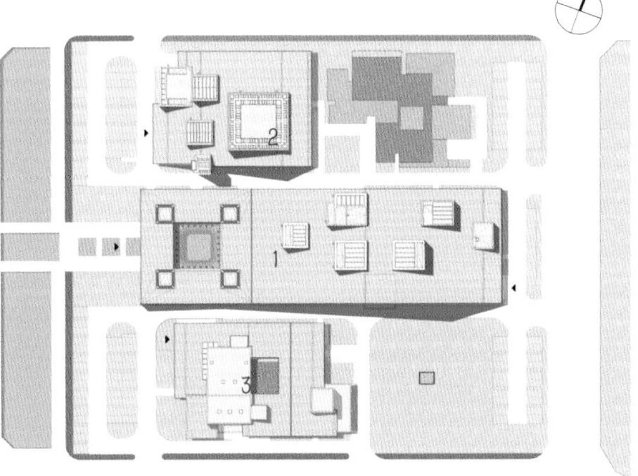

总平面图 Site plan

1 羌族民俗博物馆 Qiang Folk Museum

2 图书馆 Library

3 文化馆 Cultural Hall

建筑的形体组合来源于连绵起伏的羌族传统村寨

The building mass composition derives from the densely disposed roofs of traditional Qiang settlements.

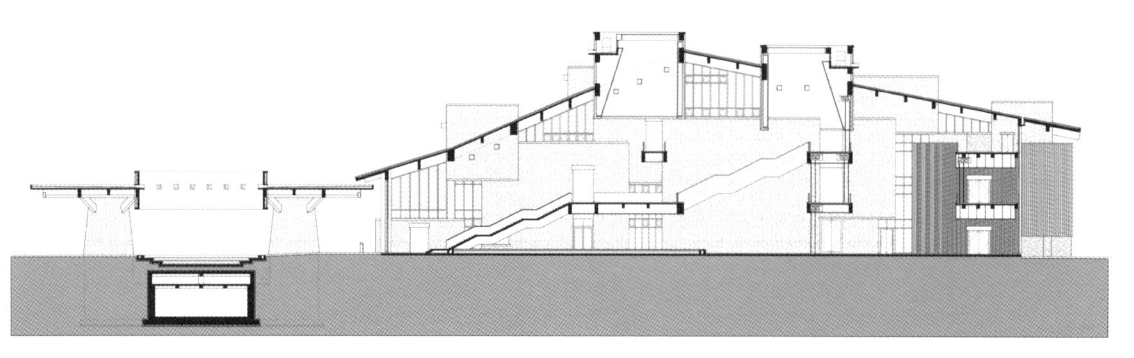

剖面图 Section

敦实厚重的前庭既提供了开敞的市民空间，也是建筑空间和体形上的重点 The open forecourt not only provides urban space for citizen activity, but also is the focus of space and for the complex.

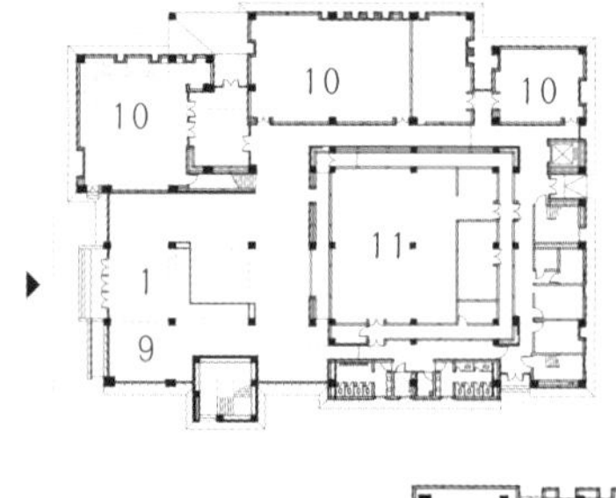

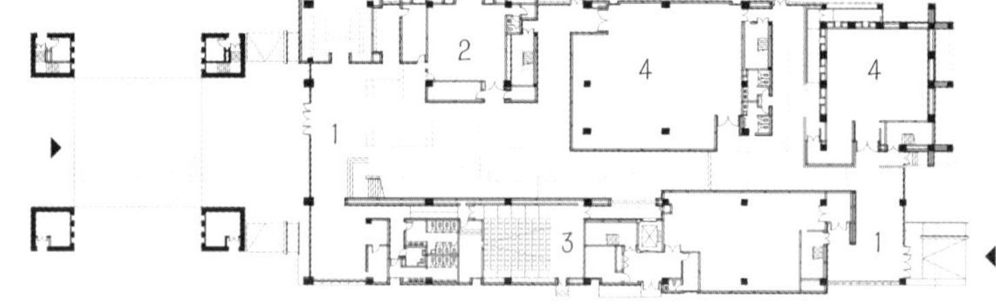

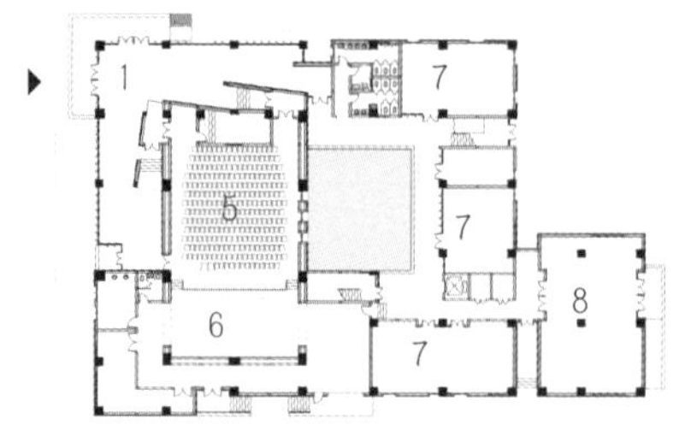

首层平面图 Ground Floor Plan

1 门厅 lobby
2 贵宾厅 VIP hall
3 报告厅 lecture hall
4 临时展厅 temporary exhibition hall
5 270座观众厅 270-seat auditorium
6 舞台 stage
7 活动室 activity room
8 展厅 exhibition hall
9 电子检索 electronic searching
10 阅览 reading room
11 特殊文献库 special document database

艺术中心 Arts Center

援　建：山东省临沂市

方　案：中国航空规划建设发展有限公司

施工图：临沂市建筑设计研究院

施　工：天元建设集团有限公司

用地面积：13 800m^2

建筑面积：10 608m^2

建筑规模：大影剧院 800座

小影剧院 各80座

Aided by:
Linyi, Shandong Province
Scheme design:
China Aviation Planning and Construction Development Co., Ltd
Construction drawings:
Planning & Architecture Design Institute of Linyi
Contractor:
Tianyuan Construction Group Co., Ltd

Site area: 13,800m^2
Building area: 10,608m^2
Capacity: large theater 800 seats
small theaters 80 seats

艺术中心位于新县城的核心地带，抗震纪念园南侧，西邻永昌河景观带，同时容纳了影剧院、川剧团、文化艺术学校等三个功能。设计从城市肌理入手，影剧院靠北设置，形成连续的沿街城市界面，强化北侧抗震纪念园的轴线感，西侧面向永昌河的公共空间则吸引市民驻足。艺术学校设于用地南部，与川剧团共同围合而成庭院空间。建筑外观简洁现代，外墙采用浅米黄色石材，按片石肌理分格，艺术墙采用白色调石材，突出现代感，三组建筑均为统一的青灰色屋顶。

The Arts Center, which is composed of the Theater, Sichuan Opera Ensemble and Art School, lie on the same plot where the core area of the new county seat is. Starting from the town texture, the design arranges the theater on the north side to form a continuous street interface. It strengthens the sense of axis of the Memorial Park on the north, and uses the public space facing the Yongchang River on the west to attract more citizens. The art school is seated on the south, enclosing a courtyard with Sichuan Opera Ensemble. Most of the exterior walls apply light beige stones, and the theme walls are covered with white stone plates to stress the modern sense. The roofs of the three buildings adopt the same grey color.

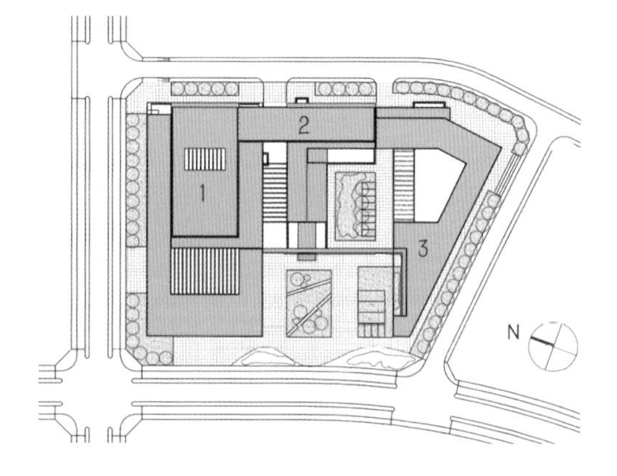

1 影剧院 Theater

2 川剧团 Sichuan Opera Ensemble

3 文化艺术学校 Art School

总平面图 Site plan

电影厅与剧场之间的半室外空间与庭院空间相连，既可将外部人流引入建筑，也加强了建筑与城镇的交融

The semi-outdoor space between the cinema hall and the theater is closely connected with the courtyard to attract external human flow into the building and strengthen communication between the building and the town.

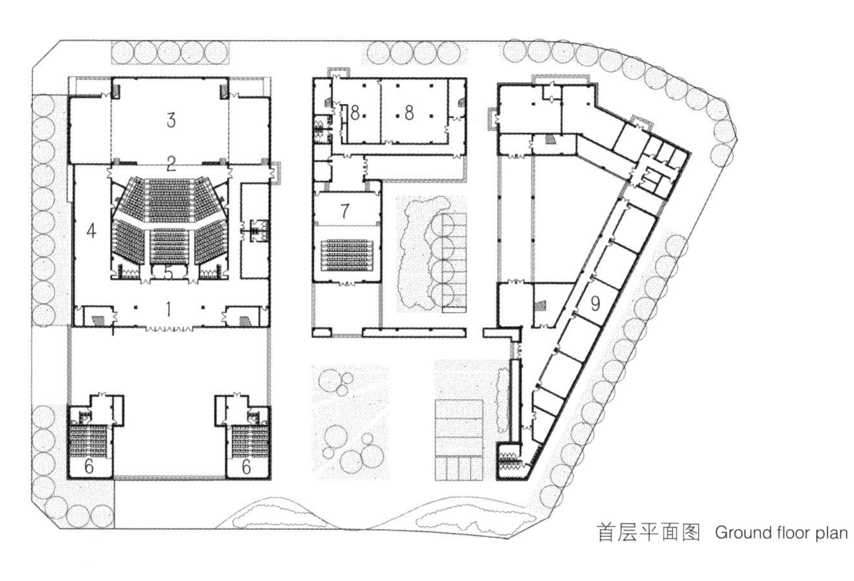

首层平面图 Ground floor plan

1 入口门厅 lobby
2 观众席 spectator seating
3 舞台 stage
4 休息厅 lounge hall
5 声光控制室 sound and light control room
6 电影厅 movie hall
7 排演厅 rehearsal hall
8 排练室 rehearsal room
9 教室 classroom

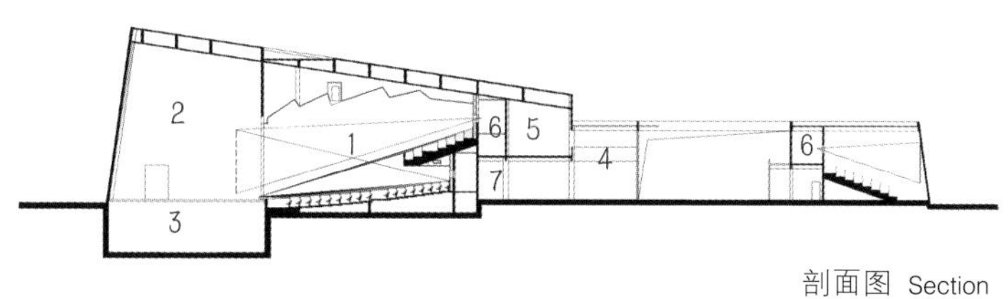

剖面图 Section

1 观众厅 auditorium
2 舞台区 stage
3 台仓 understage
4 大厅 lobby
5 休息厅 lounge Hall
6 放映室 projection Room
7 光控室 light Control Room

体育中心及青少年活动中心 Sports Center and Youth Center

援　建：山东省青岛市

设　计：青岛市建筑设计研究院

施　工：青建集团股份有限公司

莱西市建筑总公司

用地面积：57 400m²

建筑面积：20 183m²

其中　体育场 7 650m²

体育馆 6 736m²

青少年活动中心 5 797m²

建筑规模：体育场8000座

体育馆1500座

Aided by:
Qingdao, Shandong Province
Designer:
Qingdao Architectural Design & Research Institute
Contractors:
Qingjian Group Co., Ltd.
General Construction Company of Laixi

Site area: 57,400m²
Building area: 20,183m²
Stadium 7,650m²
Gymnasium 6,736m²
Youth Center 5,797m²
Capacity: Stadium 8,000 Seats
Gymnasium 1,500 Seats

体育中心及青少年活动中心位于新县城南端。用地南侧为400m跑道体育场，可承办足球、田径等多种大型体育赛事及群众集会、文艺表演，西看台为主看台，沿永昌大道的东看台做草坡处理，通过景观绿化的处理更加贴近自然。用地北侧布置体育馆和青少年活动中心。具有综合功能的体育馆可举办篮球、排球、羽毛球等室内单项比赛，满足赛前训练要求，并兼顾群众集会及演艺功能。一层主要布置比赛相关用房及运动员、记者及贵宾入口。二层是主要的观众入口层，通过疏散平台与青少年活动中心连接。青少年活动中心则是容纳体育训练、运动健身、课外活动，妇女儿童活动、青年创业实践等功能的综合场所。

Sports Center and Youth Center is situated at the south end of the new county seat. The stadium is arranged on the south of the site which can hold various large-scale events such as football, track and field matches, public assembly and art performances. The west spectator seating is the main stand, and the east stand set along the Yongchang Avenue has a green ramp, making it closer to nature through landscape greening. The gymnasium and the Youth Center are arranged on the north. The multi-functional gymnasium can hold individual events of indoor basketball, volleyball and badminton matches, satisfy the requirements of pre-match training, and hold public assembly and performances. The ground floor is mainly set with competition room and athlete, press and VIP entrances. The main spectator entrance is set on the first floor, which is connected with the Youth Center through an exit platform. During peacetime, the Youth Center serves as a venue with corresponding administrative functions for sports training, fitness, extracurricular activity, women and children activity and youth business practice.

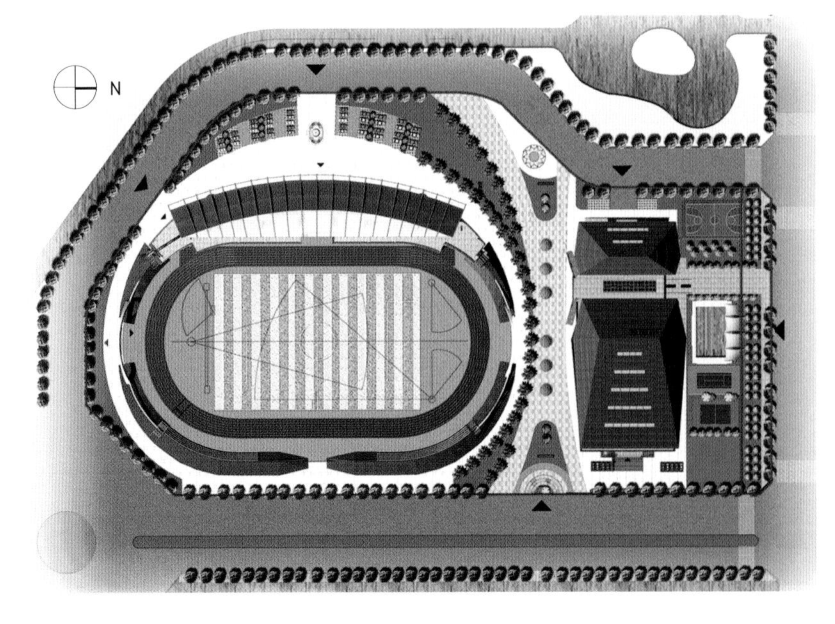

总平面图　Site plan

北川体育中心

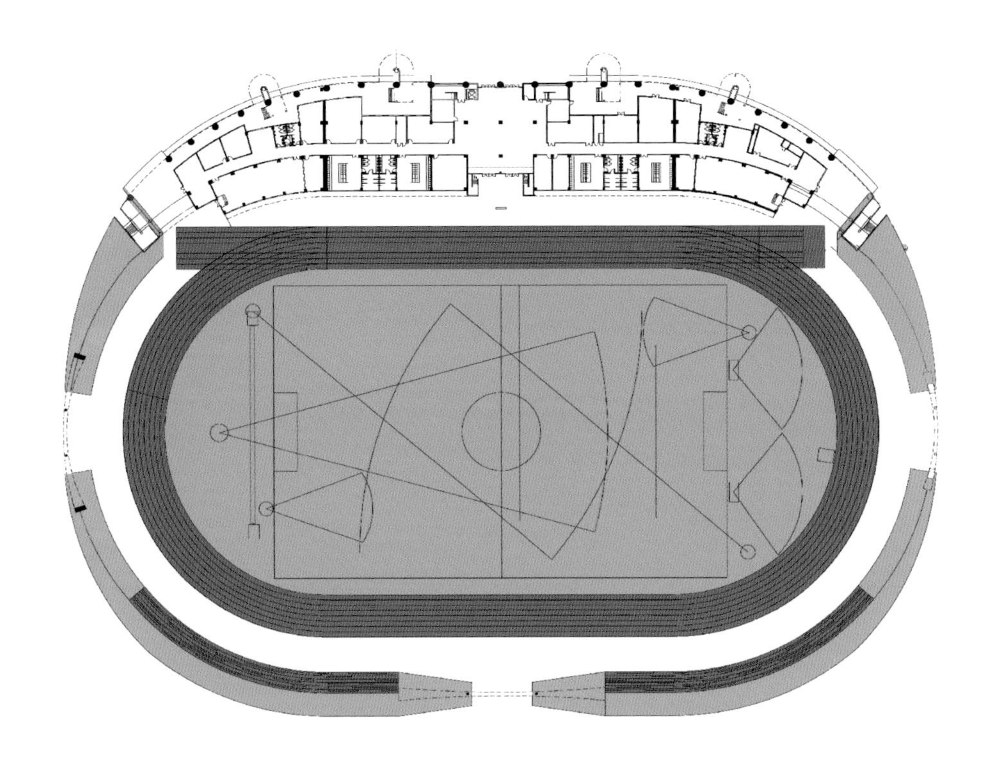

体育场首层平面图 Ground floor plan of the Stadium

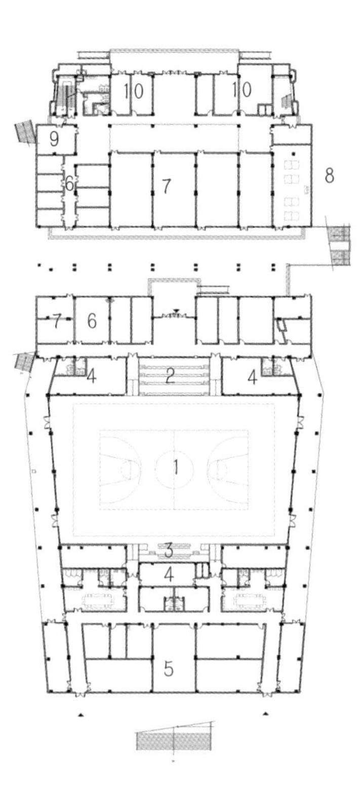

1 综合体育场 multi-functional gymnasium
2 观众席 audience seating
3 运动员席 athletes seating
4 器材库 sports equipment room
5 训练室 training room
6 琴房 piano room
7 兴趣活动室 interests activity room
8 乒乓球室 table tennis room
9 妇女儿童活动室 women and children activity room
10青年创业实践室 youth business practice room

体育馆及青少年活动中心首层平面图 Ground floor plan of the Gymnasium and Youth Center

游泳广场

北川体育中心

青少年活动中心通过二层的疏散平台与体育馆相连 The Youth Center and the Gymnasium are connected by the platform on the second floor.

行政中心 Administration Center

设计：香港华艺设计顾问（深圳）有限公司

施工：四川华西集团有限公司

用地面积：59 300m^2

建筑面积：52 000m^2

Designer:
Hongkong Huayi Designing Consultants (Shenzhen) Co., Ltd.
Contractor:
Sichuan Huaxi Group Corporation Limited

Site area: 59,300m^2
Building area: 52,000m^2

北川行政中心坐落于新县城北部山麓下，跨外环路永昌大道而设，集中布置了北川县的政府办公场所。北侧地块三面环山，建筑采用U形布局，用作党委、人民政府、人大常委会和政协委员会等四套班子的办公楼；南侧地块对称于主轴线布置了四组围合式的建筑，分别设置质检局、工商局、国税局、地税局等政府部门，以及司法局、检察院和法院等政法部门的办公楼，用地东南角隔永昌河景观带设有县档案馆。贯穿用地南北80m宽的主轴线广场，向南延伸与永昌河景观带衔接，可满足集会、庆典和市民活动的需求。

The Administration Center is situated under the northern mountains of the new county seat, and is built across the outer ring road; the north side of this project site is surrounded by mountains on three sides, where is a venue of offices for the Party committee, the people's government, the NPC and CPPCC. The south project site is used as offices for local departments directly under the people's government and political-legal departments. In addition, there is a north-south 80m-wide town square in the center of the project site, which connects the landscaped green belt along Yongchang River on the south side, and can hold assemblies and celebrations for the demands of the citizens.

通过平屋顶、坡屋顶、平坡结合的多样组合，以及围合式的台地布局，形成错落有致的天际线

Through the composition of flat roof and pitched roof as well as the enclosed stepped terrace layout, the building cluster forms a dramatic contour line.

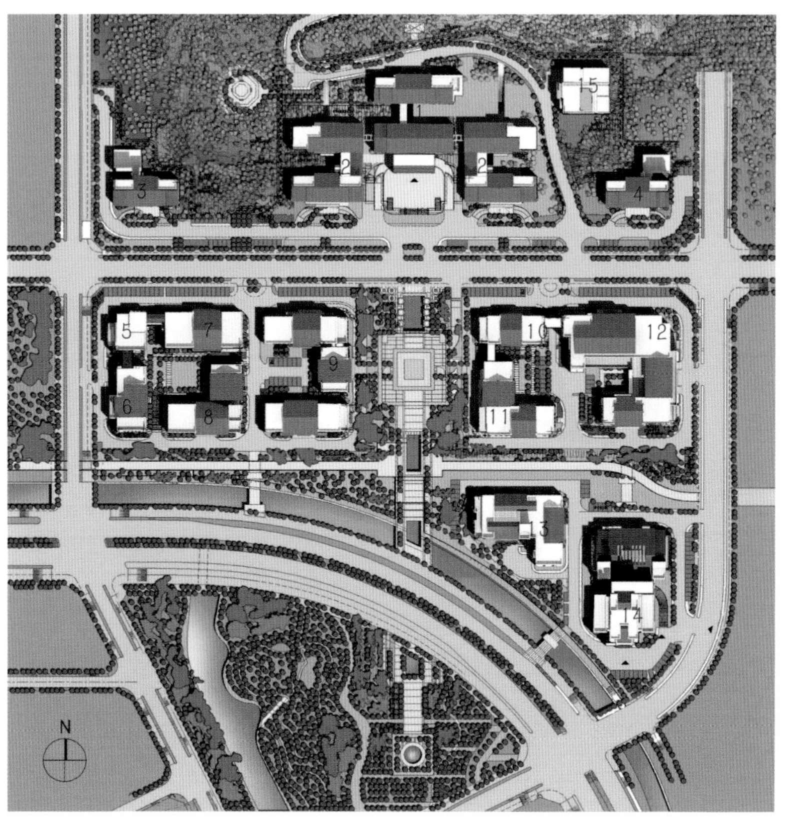

1 县委 Beichuan County Party Committee
2 县政府 Beichuan County Government
3 县人大 Beichuan County People's Congress
4 县政协 Beichuan County Political Consultative Conference
5 质监局 Quality Supervision Bureau of Beichuan County
6 工商局 Industrial and Commercial Bureau of Beichuan County
7 国税局 State Administration of Taxation of Beichuan County
8 地税局 Local Tax Bureau of Beichuan County
9 局级办公楼 local departments of Beichuan County
10 司法局 Beichuan County Justice Bureau
11 检察院 Beichuan County Procuratorate
12 法院 Beichuan County Court
13 档案馆 Beichuan County Archives
14 惠民大楼（行政服务中心）Huimin Building (Administrative Service Center)
15 食堂 dining hall

总平面图 Site plan

通过轴线、界面和群体的整合，形成横向展开的整体，体现行政建筑应有的庄重气质
The design integrates axis, boundary and building complex to form a transversely expanded integrity, and reflect solemnity of the administration office buildings.

为保证其他行政机构尽快投入使用，县委办公楼将是行政中心最后竣工的建筑

To guarantee the opening of other administrative organizations, the office building of Beichuan County Party Committee is the last completed building in the Administration Center.

惠民大楼（行政服务中心） Huimin Building (Administrative Service Center)

援　建：山东省东营市

方　案：香港华艺设计顾问（深圳）有限公司

施工图：山东信诚建筑规划设计有限公司

施　工：山东万达建安有限公司

用地面积：6 800m^2

建筑面积：11 615m^2

Aided by:
Dongying, Shandong Province

Scheme design: Hongkong Huayi Designing Consultants (Shenzhen) Co., Ltd.

Construction drawings: Shandong Xincheng Architectural Planning & Design Co., Ltd.

Contractor: Shandong Wanda Jianan Co., Ltd.

Site area: 6,800m^2
Building area: 11,615m^2

惠民大楼位于行政中心东南侧，与其隔永昌河景观带相望，集中了政务设施中面向民众的行政服务中心、惠民帮扶中心等部门以及相应的业务用房，位置相对独立，而在建筑形式上与行政中心有一定的联系。向公众开放的服务中心入口设于北侧，业务用房入口设于南侧。外形采用平屋顶与坡屋顶相结合的形式，形成高低错落的天际轮廓线，体现政府亲民形象。同时充分体现当地建筑石木结合、梁柱穿斗的特点。

The Huimin Building is situated on the south of Administrative Center and facing the Administration Center in the north side of the landscape green belt along Yongchang River. It includes the government affairs service center, people-benefited assistance center and business center, of which, the entrance of the service center that opens to the public is located on the north side, while the entrance of the business center is on the south side. It adopts a combination form of flat roof and pitched roof to form high-low scattered contour lines.

建设系统行政办公楼 Construction Department Office Building

方　案：中国建筑设计研究院

施工图：信息产业电子第十一设计研究院

施　工：四川华西集团有限公司

用地面积：16 500m²

建筑面积：7 975m²

Scheme design:
China Architecture Design & Research Group
Construction drawings:
The Eleventh Design & Research Institute of IT Co., Ltd.
Contractor:
Sichuan Huaxi Group Corporation Limited

Site area: 16,500m²
Building area: 7,975m²

建设系统行政办公楼位于北川新县城北部，城市外环路的北侧，集中了北川羌族自治县地震局、国土局、环保局、建设局、房管局等五个政府部门的办公场所。基地背靠自然山体，面对北川中学，高差变化较大。设计结合用地高差以及多单位共存等特征，通过8个3层的L形建筑单体，围合成两组院落，构成了完整的建筑群落，人们可从南侧入口拾级而上，进入内院。设计营造出舒适安宁的办公组团氛围，并通过灰白与羌红的色调搭配以及高差变化的天际线设计，共同体现了羌族文化的传统特征。

This project is located on the north side of the new county seat, which is against natural mountains and faces Beichuan High School. Considering the large altitude difference of the site and the juxtaposition of five administrative departments, Seismological Bureau, Land and Resources Bureau, Environmental Protection Bureau, Construction Bureau, and Realty Administrative Bureau, the design adopts eight L-shaped buildings with three stepped terraces to compose two courtyards and form an orderly building complex. Therefore, people can ascend the stairs from the south entrance, and enter the inner courtyard. It also obtains a comfortable and peaceful atmosphere, and adopts a palette of grey and red and fluctuated skyline to embody the traditional architecture features of Qiang nationality.

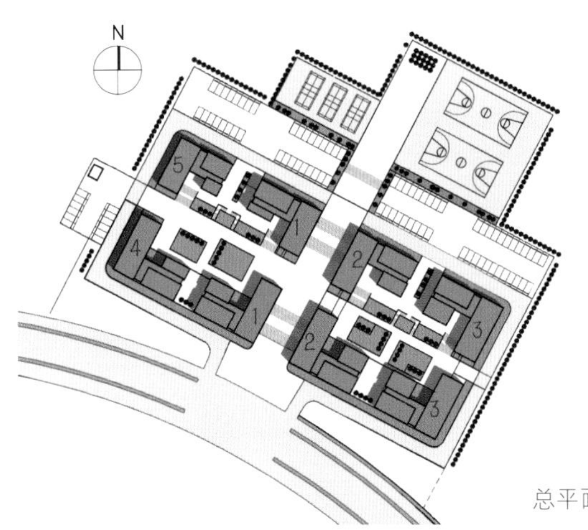

1 国土局
the Land and Resources Bureau
2 建设局
the Construction Bureau
3 地震局
the Seismological Bureau
4 环保局
the Environmental Protection Bureau
5 房管局
the Realty Administrative Bureau

总平面图 Site plan

办公组团贴近街道布置，强化了建筑在城镇中的形象氛围
The office buildings are arranged near the street, enhancing the image of the building in the town.

简洁明快的体块组合，在山麓间形成高低错落的轮廓线

Concise mass combination forms high-low scattered contour lines among mountains.

围合式的组团布局形成安静的内部办公环境　With the enclosed layout, a quiet interior office atmosphere is generated.

农业系统行政办公楼 Agricultural Department Office Building

方　案：中国建筑设计研究院

施工图：信息产业电子第十一设计研究院

施　工：四川华西集团有限公司

用地面积：10 346m²

建筑面积：10 400m²

Scheme design:

China Architecture Design & Research Group

Construction drawings:

The Eleventh Design & Research Institute of IT Co., Ltd.

Contractor:

Sichuan Huaxi Group Corporation Limited

Site area: 10,346m²

Building area: 10,400m²

农业系统行政办公楼位于北川新县城东部，将北川羌族自治县林业局、农业局、水务农机局/开茂水库工程建设管理局、畜牧兽医管理局等四个政府部门的办公场所整合为一体。四个政府部门各自占据地块一角，围合出公共交流庭院，形成有机完整的办公组团。体量最大的林业局布置在邻近道路交叉口的西南角，作为主要展示形象，体量次之的农业局和水务局分置西北和东南角，体量最小的畜牧局设置于东北向，便于引入自然景观。建筑以简洁的横竖线条构筑完整的群体形象，并注重尺度、材质和色调的协调，同时又根据四个部门的行业特点，在立面划分上体现了森林的竖向生长，田地的规则四方，水流的横向延伸和动物四散灵动的自由形态。

This project site is located on the east side of the new county seat. The four governmental departments are designed and integrated as a whole, with each occupying a corner of the site, which enclose a courtyard for public communication and form an organic and integrative office group. The Forestry Bureau with the maximum size is set in the southwest corner near the road intersection as a main image; the Agriculture Bureau and Water Affairs Bureau are respectively set in the northwest and southeast corners; in consideration of introducing natural landscape in the north-east direction, the Animal Husbandry Bureau with the minimum size is set in the northeast corner. It forms an integrative image with concise and forceful horizontal and vertical lines, and focuses on coordination of scale, material and color tune. The façade of each bureau also reflects its own industry characteristics: vertical growth of woods for the forestry bureau, a regular and square shape for the agriculture bureau, a transverse extension of water flow the water bureau and a free status of animals for the animal husbandry bureau.

1　林业局 the Forestry Bureau
2　农业局 the Agriculture Bureau
3　水务局 the Water Affairs Bureau
4　畜牧局 the Animal Husbandry Bureau

社会保障中心及卫生系统办公楼 Social Security Center & Health Department Office Building

援　建：山东省莱芜市

方　案：中国工程物理研究院建筑设计院

施工图：山东中大建筑设计院有限公司

施　工：山东莱芜建设集团有限公司

　　　　山东正顺建设集团有限公司

用地面积：10 800m²

建筑面积：11 198m²

Aided by:
Laiwu, Shandong Province
Scheme Design:
Architectural Design Institute of CAEP
Construction drawings:
Shandong Zhongda Architecture Design Institute Co., Ltd.
Contractors:
Shandong Laiwu Construction Group Co., Ltd
Shandong Zhengshun Construction Group Co., Ltd

Site area: 10,800m²
Building area: 11,198m²

该地块位于新县城东北部，主要容纳四个功能：就业和社会保障综合服务中心、人口和计划生育服务大楼、卫生综合大楼和卫生局进修楼。基地位于新县城北部，采用围合式布局，红十字会仓库相对独立，设于地块中央，自然分隔出分属社保中心和卫生系统的庭院空间。社保中心和计划生育服务楼服务市民的入口均沿街设置，办公入口则设于内院中，避免了相互干扰。建筑的高度变化和功能块的相互组合，形成了错落有致的外部空间。

The project is located on the northeast part of the new county seat, which is composed of four buildings: the Employment and Social Security Comprehensive Service Center, the Population and Family Planning Service Building, the Health Complex Building and the Health Vocational Studies School. The site is located on the north side of the new county seat and adopts an enclosed layout. The Red Cross Warehouse is set in the middle of the plot, naturally separating the courtyard space for the Social Security Comprehensive Service Center and the Health Bureau. The public entrances of the Social Security Comprehensive Service Center and the Population and Family Planning Service Building are arranged along the street, and the internal entrances are set in the inner courtyard to avoid mutual interference. The building height changes and combination of functional blocks form an orderly arrangement of the exterior space.

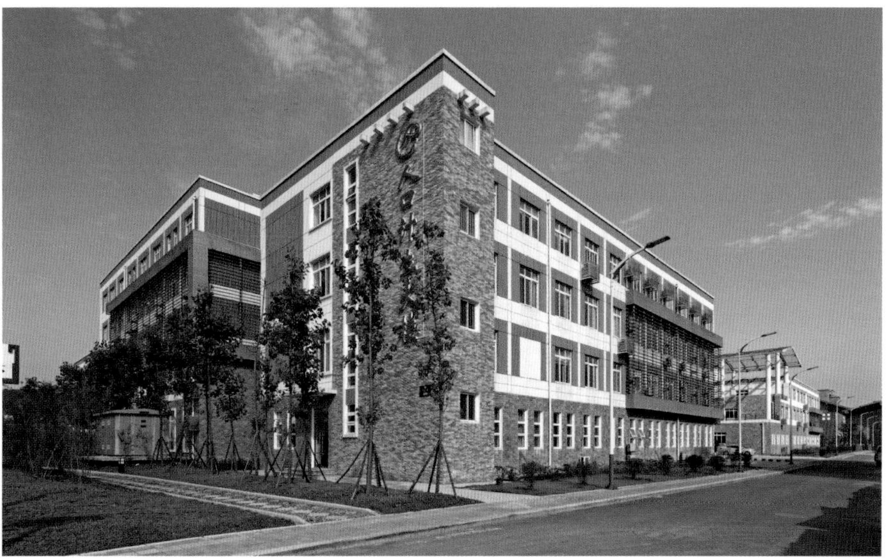

1 就业和社会保障综合服务中心
Employment and Social Security Comprehensive Service Center
2 计划生育指导站
Population and Family Planning Service Building
3 卫生局办公楼
Office Building of the Health Bureau
4 爱国卫生运动委员会办公室
Office of Patriotic Health Campaign Committee
5 红十字仓库及卫生进修学校
Red Cross Warehouse and Health Vocational Studies School

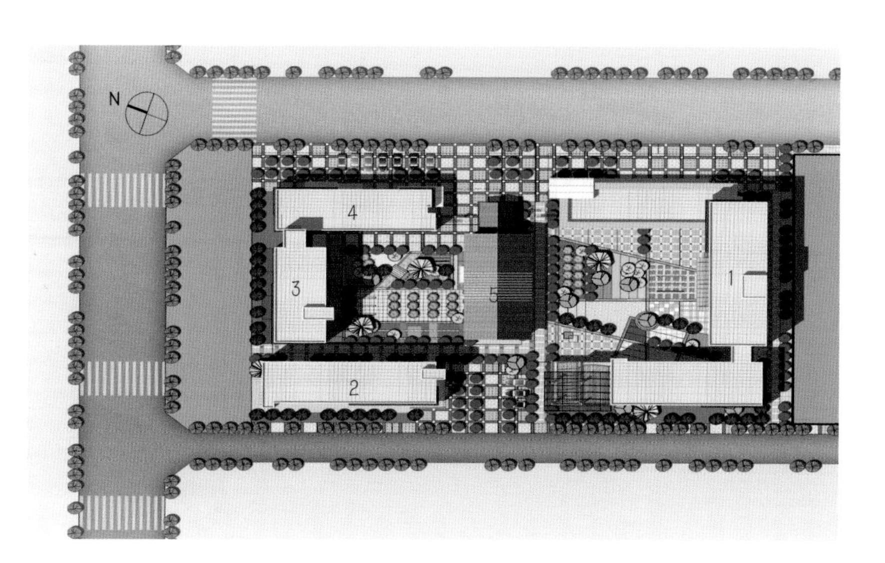

总平面图 Site plan

公安局 Public Security Bureau

设　计：中国建筑标准设计研究院

施　工：四川华西集团有限公司

用地面积：10 000m²

建筑面积：12 350m²

Designer:
China Institute of Building Standard Design & Research

Contractor:
Sichuan Huaxi Group Corporation Limited

Site area: 10,000m²

Building area: 12,350m²

公安局综合楼位于新县城东部，环路外侧，由公安局办公楼和县应急中心组成。平面布局为U形，由三部分组成，6层的公安局办公楼位于中部，其南侧为主入口广场，东为公安局对外办公部分，3层的应急指挥中心则位于用地西部，通过走廊与办公楼相连。主入口广场位于用地南侧，建筑围合的内院可停放公安局及应急指挥中心车辆，实现人车分流。

The Public Security Bureau is located on the east part of the new county seat, and is composed of the Public Security Bureau office building and the Emergency Command Center. In a U-shaped layout, the six-storey office building lies on the south of the project site, with an entrance square in front of the building; the three-storey Emergency Command Center is on the west side of the project site. The east side of the project site is a citizen service department of Public Security Bureau, while the south side is a main entrance square to the compound. Thus the inner courtyard enclosed can be used as a path for vehicle to separate pedestrians away from vehicle.

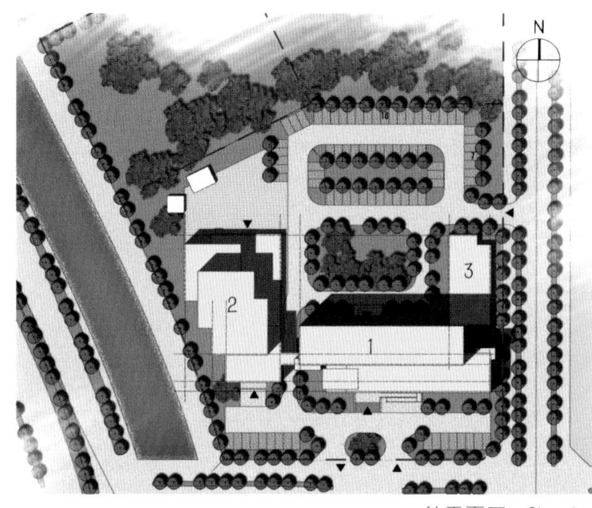

总平面图 Site plan

1 公安局办公楼 Public Security Bureau Office Building
2 应急指挥中心 Emergency Command Center
3 公安局对外办公 Citizen Service Department

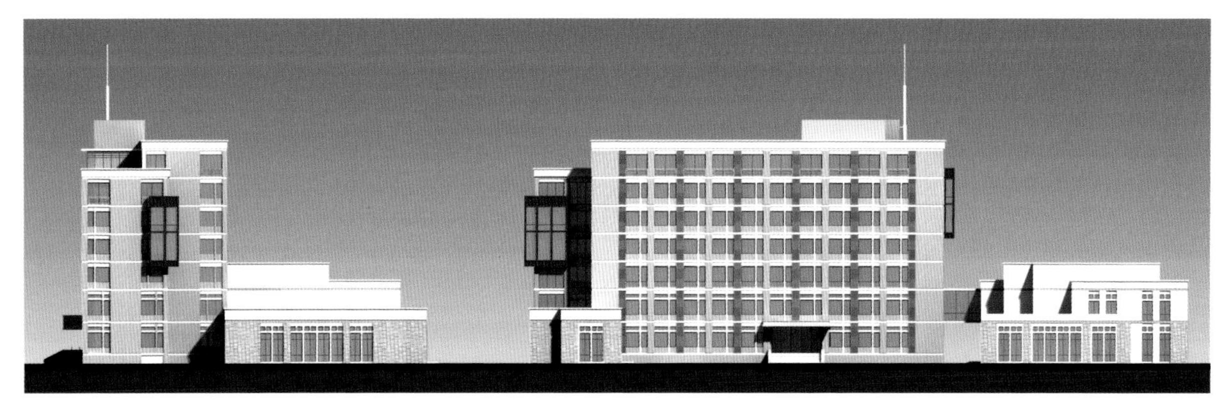

立面图 Elevation

30

交警大队 Traffic Police Station

方　案：中国建筑标准设计研究院

施工图：信息产业电子第十一设计研究院

施　工：四川华西集团有限公司

用地面积：10 000m²

建筑面积：2 796m²

Scheme design:
China Institute of Building Standard Design & Research
Construction drawings:
The Eleventh Design & Research Institute of IT Co., Ltd.
Contractor:
Sichuan Huaxi Group Corporation Limited

Site area: 10,000m²
Building area: 2,796m²

公安局交警大队、车管所和永昌中队设于同一地块内，位于新县城东端。交警大队作为主要建筑，面向南侧主要城市道路，永昌中队和车管所则面向东侧街道，其后围合出庭院空间及摩托车检验、考试场地。L形的布局不仅形成了完整而清晰的城市界面，也突出了建筑的整体形象。三栋建筑通过连廊相互连接，在外观上通过材料、色彩以及体块的错落咬合，体现了传统地域特点。

Buildings on this project site include Traffic Police Station, Vehicle Administrative Office and Yongchang Traffic Police Squadron. The whole site is situated on the east part of the new county seat, of which the Traffic Police Station is set beside the main road on the south side of the site, and the Yongchang Traffic Police Squadron and the Vehicle Administrative Office face the east street. Such an L-shaped layout acquires a complete and clear town interface, and also highlights the overall image of these buildings. Through connection of corridors, these three buildings enclose a courtyard and places for motorcycle test and examination.

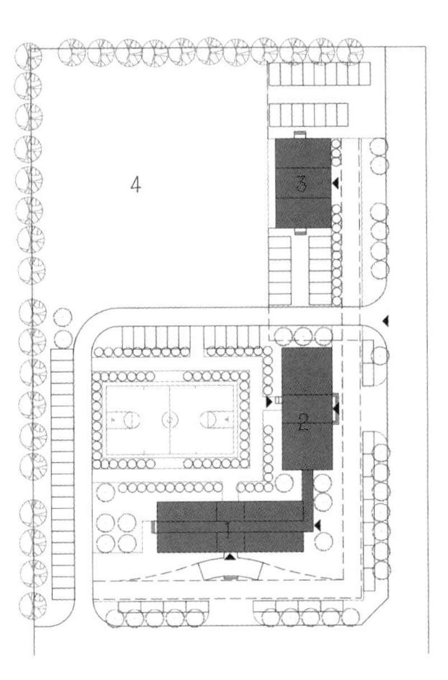

1 交警大队
Traffic Police Station
2 永昌中队
Yongchang Traffic Police Squadron
3 车管所
Vehicle Administrative Office
4 摩托车检验场
motorcycle test field

总平面图　Site plan

交通警察
TRAFFIC POLICE

消防站 Fire Station

方　案：中国建筑设计研究院

施工图：信息产业电子第十一设计研究院

施　工：四川华西集团有限公司

用地面积：10 382m²

建筑面积：4 171m²

Scheme design:
China Architecture Design & Research Group
Construction drawings:
The Eleventh Design & Research Institute of IT Co., Ltd.
Contractor:
Sichuan Huaxi Group Corporation Limited

Site area: 10,382m²
Building area: 4,171m²

公安局消防大队位于新县城南部，因需设置南北向训练跑道，由营房和训练塔组成的主体建筑亦南北向直线布置，设于用地东侧，西侧则为训练场。车库布置在营房首层，面向城市道路，便于快速出车。二、三层分别布置战士宿舍和办公用房。建筑体形在一定程度上模仿了周边自然山石的形态，体现消防队员的顽强精神，也符合羌族传统建筑坚固且错落的特点。

The Fire Brigade is located on the south side of the new county seat, and its main buildings consist of barracks and training tower, which are separately set on the east side of the site, while the west side is designed as training grounds and an inner courtyard. The entrance of the Fire Brigade is set along the main road of the town, with its garage exactly facing the road, which is convenient for quickly dispatching vehicles. The second floor and third floor are firefighters' dormitories and offices. Its appearance imitates rock forms to embody indomitable will of fire fighters, and also conforms to the traditional Qiang nationality building features of firmness and free distribution.

人民武装部 People's Melitia Department

设　计：中国建筑标准设计研究院

施　工：绵阳市方圆建筑工程公司

用地面积：10 499m^2

建筑面积：7 766m^2

Designer:
China Institute of Building Standard Design & Research
Contractor:
Mianyang Fangyuan Construction Co., Ltd.

Site plan: 10,499m^2
Building area: 7,766m^2

人民武装部位于新县城北部山麓，背靠自然山体，场地内高差较大。主入口位于用地南侧，设计上庄重严肃，正对北侧办公楼，院内设有队列训练场。办公楼东北角设干部周转房，较为安静。民兵训练基地设于用地西南角，平面为L形，用地西北角则为民兵训练场地。建筑由简洁明确的体块构成，凹凸有致的处理手法使立面产生丰富的光影效果。坡屋顶对当地穿斗木构做法的抽象表达体现了建筑的地域特色。

The People's Melitia Department is situated on the foot of the north mountain of the new county seat with a large height difference. The main entrance is located on the south side, and faces the office building on the north side. A training field is set in the compound. The temporary housing building for cadres is set on the northeast corner of the office building. An L-shaped militia training base is arranged in the southwest corner, with its training field in the northwest corner of the site. The building mass is concise, and its façade generates rich light and shadow effect. The pitched roof demonstrates the local feature of architecture.

县委党校 Beichuan CCP School

方　案：中国建筑设计研究院

施工图：信息产业电子第十一设计研究院

施　工：四川华西集团有限公司

用地面积：7 800m^2

建筑面积：2 458m^2

Scheme design:
China Architecture Design & Research Group
Construction drawings:
The Eleventh Design & Research Institute of IT Co., Ltd.
Contractor:
Sichuan Huaxi Group Corporation Limited

Site area: 7,800m^2
Building area: 2,458m^2

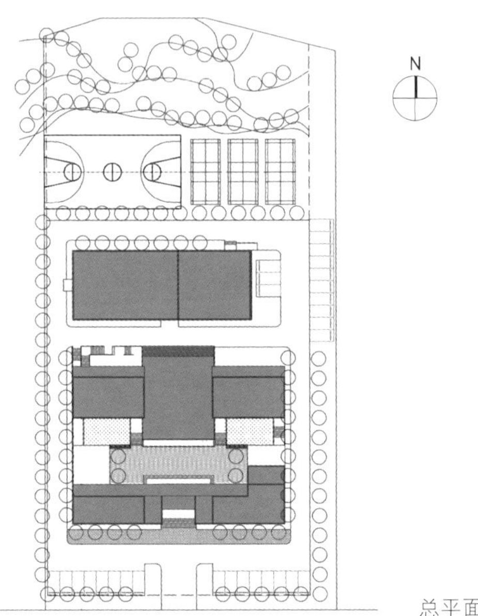

总平面图
Site plan

党校地处北川新县城北部，外环路北侧，背靠自然山体，拥有良好的自然氛围，毗邻人民武装部和建设系统办公楼。设计结合用地较大的高差，依山就势，将建筑分为三组平行于街道的单体，依次为办公用房，教学及会议用房，以及住宿餐饮用房，吸取羌寨依山就势的特点，层层高起，通过天际线的变化和材料、色彩的继承，体现了羌族传统特征。

The CCP School is situated in the north part of the new county seat, to the north side of the outer ring road, and is close to natural mountains, which has favorable natural environment with changing height of the land. According to the large altitude difference, the project is designed as three parallel building masses, containing the offices, classrooms and living service rooms. Through the variation of skyline and application of the local materials, it embodies traditional features of the Qiang nationality culture.

教师进修学校 Teachers' Training School

设　计：北京国城建筑设计公司

施　工：四川华西集团有限公司

用地面积：4 992m²

建筑面积：3 913m²

Designer:
China Academy of Urban Planning & Design
Beijing Guocheng Architectural Design Company Branch
Contractor:
Sichuan Huaxi Group Corporation Limited

Site area: 4,992m²
Building area: 3,913m²

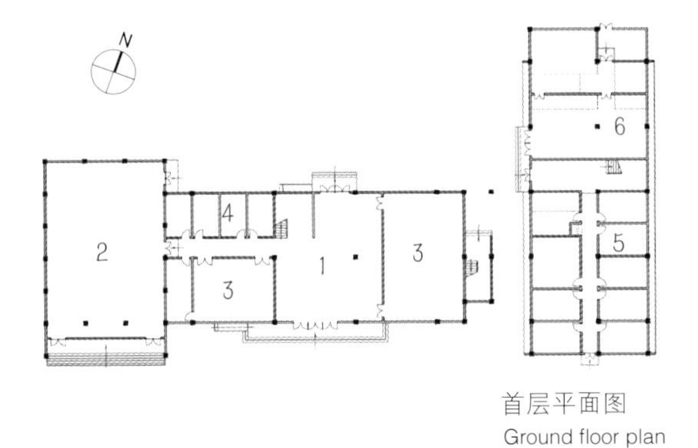

首层平面图
Ground floor plan

1 门厅 foyer
2 多功能厅 multi-functional hall
3 教室 classroom
4 办公室 office
5 宿舍 dormitory
6 食堂 dining hall

教师进修学校靠近新县城主轴线东段，南邻北川文化中心。设计力求在布局和体量上与文化中心相协调。平面呈L形布局，沿街道展开，南侧部分为教学楼，东侧为宿舍楼。立面设计吸收当地羌族建筑木石结合的特点，并通过材料和建筑语汇的统一，起到对周围主要建筑的烘托作用。

The Teachers' Training School is located nearby the cultural axis and to the south of the Cultural Center, with its design emphasized on coordinating with the Cultural Center in the layout and size. Adoption of wood and stone in the facade, application of the Qiang architectural features and combination with the local context features play important roles of setting off the surrounding landmarks.

永昌镇人民政府 The People's Government of Yongchang Town

方　案：深圳市建筑设计研究总院有限公司

施工图：信息产业电子第十一设计研究院有限公司

施　工：四川华西集团有限公司

用地面积：2 500m²

建筑面积：1 743m²

Scheme design:
Shenzhen General Institute of Architectural Design and Research Co., Ltd.
Construction drawings:
the Eleventh Design & Research Institute of IT Co., Ltd.
Contractor:
Sichuan Huaxi Group Corporation Limited

Site area: 2,500m²
Building area: 1,743m²

永昌镇作为北川新县城所在地，有其自身的镇政府职能。镇政府办公楼设于新县城主轴线东端，北邻北川文化中心。建筑由两部分组成，北为公共服务大厅及办公楼，南为会议部分，二者通过廊架联系。建筑体块、三段式划分、坡屋顶、挑台等方面传统语汇的使用，暖灰色石材、白色涂料、仿木色构件，以及传统样式花窗、木质阳台、连廊、门楣和窗楣等细部的刻画，都加强了建筑的传统元素表达。

As a venue for the new county seat, the Yongchang Town has its own township government functions. The office building of the government is located on the east end of the cultural axis, and to the north of the Cultural Center. The building is composed of two parts: the public service hall and office building on the north side and meeting rooms on the south, which are connected by a gallery. Traditional architectural elements are expressly demonstrated by using building mass, ternary form, pitched roofs, cantilever balcony, warm grey stones, white coating, timber-like components, lattice windows, wooden balconies, connecting gallery, door lintels and window lintels, etc.

N

首层平面图
Ground floor plan

中国共产党北川羌族自治县永昌镇委员会
中国共产党北川羌族自治县永昌镇纪律检查委员会

广播电视中心 Radio and Television Center

援　建：山东省济南市

设　计：山东同圆设计集团有限公司

施　工：山东平安建设集团有限公司

用地面积：2 900m²

建筑面积：3 507m²

Aided by:
Jinan, Shandong province
Designer:
Shandong Tongyuan Design Group Co., Ltd
Contractor:
Shandong Ping'an Construction Group Co., Ltd

Site area: 2,900m²
Building area: 3,507m²

广播电视中心位于新县城中心区北部，永昌河景观带和永昌大道之间。建筑布局呈L形，5层的办公部分南北向布置，位于基地南侧，对外功能部分则沿永昌大道布置，二者共同围合出一个面向西侧湖面的公共广场。建筑造型来自对羌族建筑原型的提炼，以碉楼的形式作为建筑物突出的主题，通过对出挑的吊脚楼、错落的屋顶平台、室外楼梯、过街楼等羌族特色元素进行抽象、变形，形成建筑体形，并在细部处理上采用了羌族装饰图案。

The Radio and Television Center is located on the north part of the core of the new county seat and between the landscaped green belt along Yongchang River and Yongchang Avenue. In an L-shaped layout, it has a five-storey office building on the south. Its external functional area is arranged along the Yongchang Avenue. They enclose a public square facing the lake on the west. The building mass is a purification of the traditional Qiang building, adopting turret as the theme of the building. The basic building shape is formed by abstracting and changing the form of the distinctive Qiang elements such as overhanging stilted building, roof platform, external stairway, arcade and Qiang decorative patterns in details.

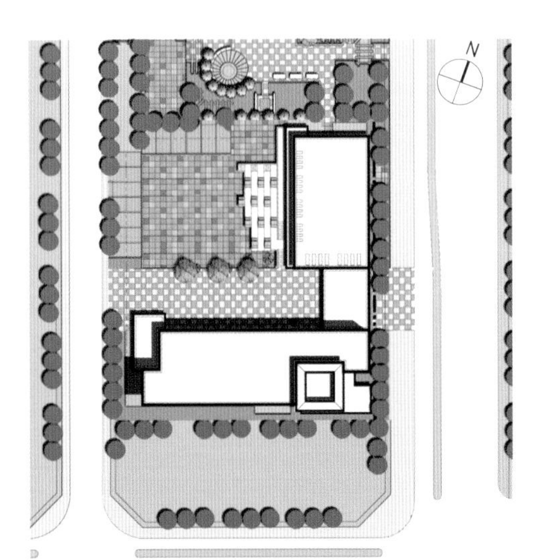

总平面图
Site plan

转播塔的比例和收分，将广电中心的功能需求与羌族传统碉楼的形式结合在一起

The proportion and tapering shape of the broadcasting tower make a harmonious combination of functional requirements and the symbolic feature of traditional Qiang turret.

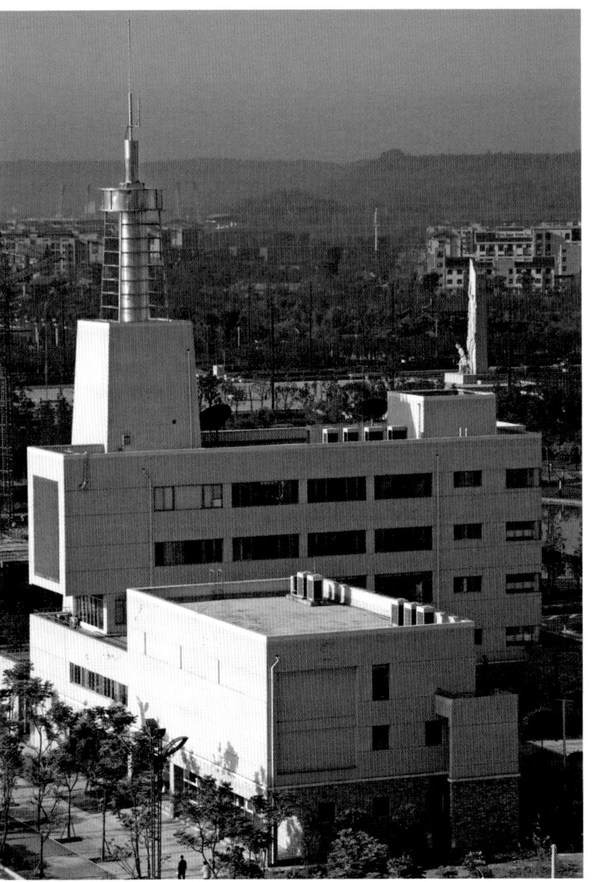

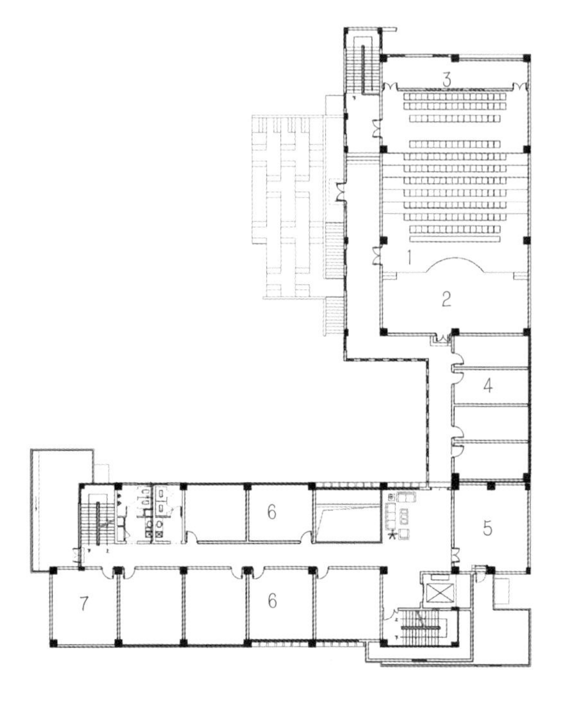

1 180人转播大厅 180-seat broadcasting hall
2 舞台 stage
3 控制室 control room
4 化妆间 dressing room
5 实况转播室 live broadcasting room
6 制作间 production room
7 录影棚 video studio

二层平面图 Second floor plan

气象站 Meteorological Station

方　案：清华大学建筑设计研究院

施工图：四川海辰工程设计研究有限公司

施　工：四川华西集团有限公司

用地面积：13 300m²

建筑面积：1 836m²

Scheme design:
Architectural Design and Research Institute of Tsinghua University
Construction drawings:
Sichuan Haichen Engineering Design Co., Ltd.
Contractor:
Sichuan Huaxi Group Corporation Limited

Site area: 13,300m²
Building area: 1,836m²

气象站位于新县城城区外东北部山麓，永昌大道西侧。设计从地段所在的坡地环境出发，吸收羌寨依山而建的特点，顺应山势形成自上而下、层层跌落的平台系统，最高层是按气象专业要求设置的观测广场，其下建筑体量随高程变化而逐级递降，结合竖向交通流线，形成一系列屋顶平台。建筑根据气象局功能的需求，以对外办公、对内办公和辅助用房三个组团，通过小单元的相互组合，形成丰富而有机的聚落形态，以现代的方式效法羌族民居小体量组合形成的群落式空间特点。

The Meteorological Station is located northeast to the new county seat. Its design starts from its gradient environment, absorbs the Qiang village features against mountains, and forms a layer by layer falling platform system from top to bottom along with mountains, of which the highest level is designed as an observation square required by the meteorological specialty. The buildings below descend along the height variation, and form roof platforms in a combination of the vertical traffic stream lines. According to the requirements of meteorological station, it is composed of external offices, internal offices and auxiliary rooms in three groups, which forms an abundant and organic settlement pattern of small units, and also shape into a small Qiang house with modern design methods.

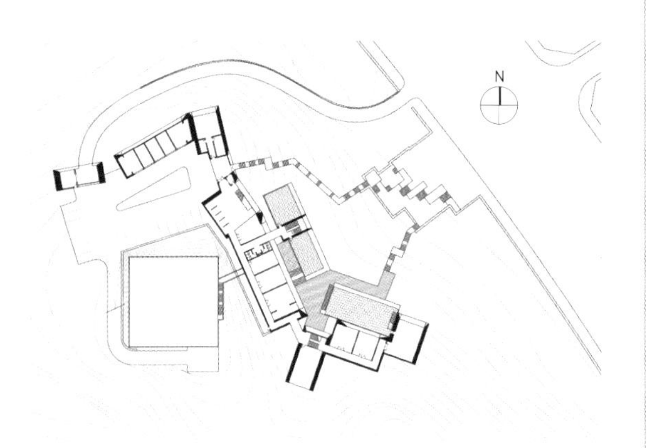

入口层平面图 Entrance floor plan

新北川宾馆 New Beichuan Hotel

援　建：山东省淄博市

方　案：青岛建筑设计研究院股份有限公司

施工图：淄博市建筑设计研究院

施　工：山东天齐置业集团股份有限公司

用地面积：14 700m²

建筑面积：20 016m²

Aided by:
Zibo, Shandong province
Scheme design:
Qingdao Architectural Design & Research Institute Co., Ltd
Construction drawings:
Zibo Architectural Design & Research Institute
Contractor:
Shandong Tanki Industry & Commerce (Group) Co., Ltd

Site area: 14,700m²
Building area: 20,016m²

新北川宾馆处于北川新县城的核心地段，抗震纪念园的北侧。设计因而引入两条轴线，强化主入口向心性的东向轴线，和尊重纪念园的南向次入口轴线。宾馆首层南侧为餐饮部分，设有两个庭院，北侧围绕景观庭院设置酒吧、咖啡馆、茶室、泳池等休闲设施，其上二到五层均为客房，通过不同部位的退台，形成景观良好的屋顶花园，并结合平顶和坡顶的交织，形成丰富的沿街立面。墙面以浅米色、白色涂料为主，局部装饰棕黄色，屋顶用青灰色屋面瓦。栏杆、百叶窗等带有羌族文化符号，同时通过有韵律的立面元素和大面积玻璃幕墙，以现代材料和设计手法演绎了传统的汉羌风格。

The New Beichuan Hotel is located at the core area of the new county seat, and faces the Memorial Park to the south. Two axes are introduced in the design, stressing the centripetal main entrance axis in the south while respecting the secondary entrance axis in the south of the Memorial Park. On the ground floor of the hotel, a dining area is set on the south with two courtyards; bar, coffee, tea house, swimming pool and other recreation facilities are set on the south centering on a landscape yard. Guest rooms are arranged on the first to the fourth floors, and a roof garden with beautiful landscape is formed through different setbacks of the top. With the arrangement of flat roof and pitched roof, diversified street façades are formed. Walls are mainly painted with light beige and white coating, which are decorated with brown coating. The pitched roofs are covered with grey tiles. Decorated with the Qiang cultural symbols, the railings and window lattices represent the local Han-Qiang architectural features with modern materials and design methods such as the large-area glass curtain walls and the rhythmic façade composition.

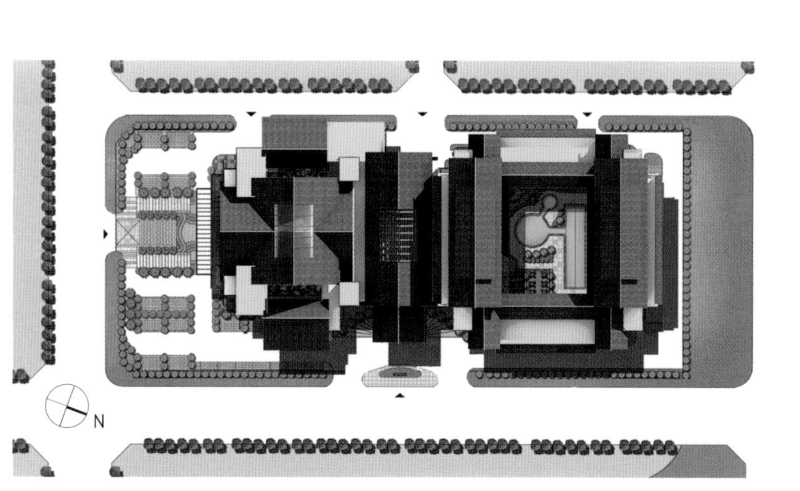

总平面图 Site plan

新北川賓館
NEW BEICHUAN HOTEL

新北川賓館

大堂装饰突出对羌族文化的表达，运用大尺度的处理手法，并结合齐鲁文化的建筑形式

Large-scale method is applied for the decoration of the lobby to highlight the Qiang culture and incorporate with the architectural style of Shandong culture.

1 大堂 lobby
2 服务总台 reception desk
3 总台办公 storage front desk
4 休息区 lounge area
5 商务中心 commercial center
6 餐厅 dining hall
7 包间 private room
8 厨房 kitchen
9 茶室 teahouse
10 游泳池 swimming pool

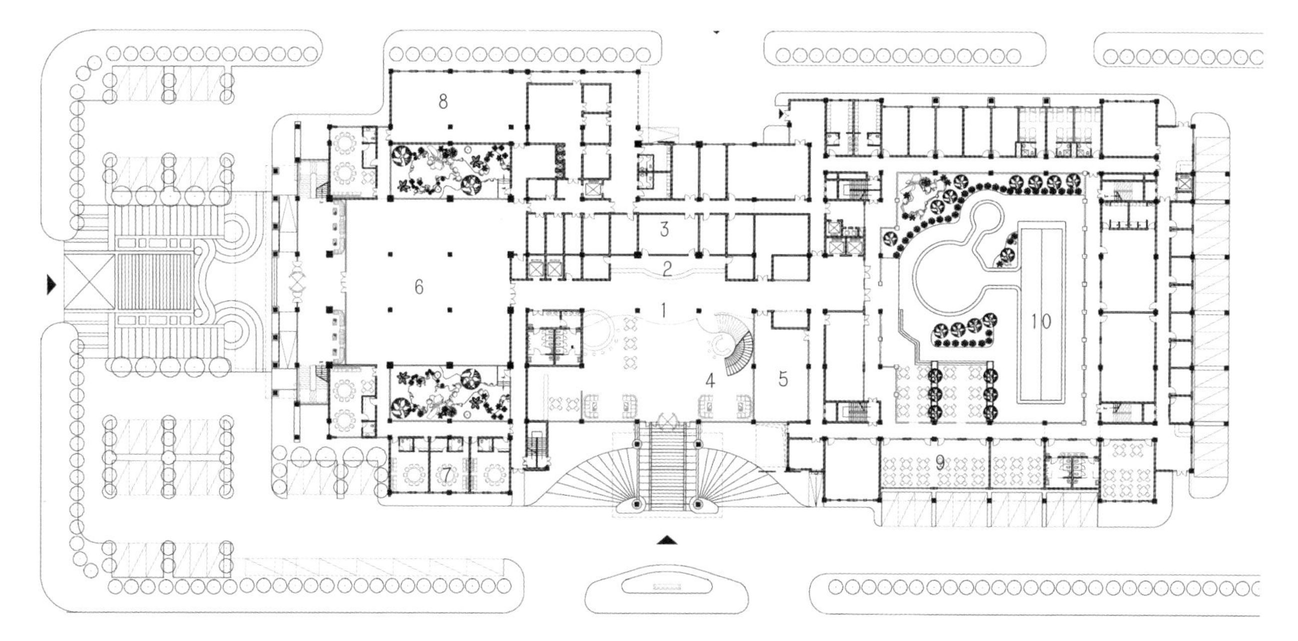

首层平面图 Ground floor plan

宾馆主入口面向东侧的永昌河公园景观带，可将景观引入位于宾馆中部的大堂，形成东西向的宾馆主轴线

The main entrance of the hotel faces the landscaped green belt along Yongchang River in the east, which can bring the landscape into the lobby and form a main east-west axis of the hotel.

央企和金融机构办公楼 Central Enterprises & Financial Organization Office Building

设　计：中科院建筑设计研究院有限公司

施　工：四川华西集团有限公司

用地面积：24 680m^2

建筑面积：40 305m^2

Designer:
Institute of Architecture Design and Research, Chinese Academy of Sciences
Contractor:
Sichuan Huaxi Group Corporation Limited

Site area: 24,680m^2
Building area: 40,305m^2

企业集群央企和金融机构办公楼位于新县城核心区北部，跨永昌大道而设，南邻纪念园和文化中心，共由9座单体建筑组成，容纳了14家邮政、银行、保险、电信等方面与人们生活密切相关的服务性企业。建筑设计追求简约、规则，竖向划分的石材分格和较为自由的开窗形式赋予立面统一而有变化的效果，并且形成了完整而富于秩序感的城市界面。

Located on the north side of the core area of the new county seat, the Central Enterprises & Financial Organization Office Building are built across the Yongchang Avenue and adjacent to the Memorial Park and the Cultural Center in the south. The cluster is composed of 9 single buildings, in which 14 enterprises covering postal, bank, insurance, telecom and other services closely involved in the citizens' life. The design for these buildings pursues concise and regular forms by applying vertical stone divisions and free windowing to endow the façades with unified and changing effects, and form a complete city interface with a sense of order.

ICBC
中国工商银行
ICBC 中国工商银行

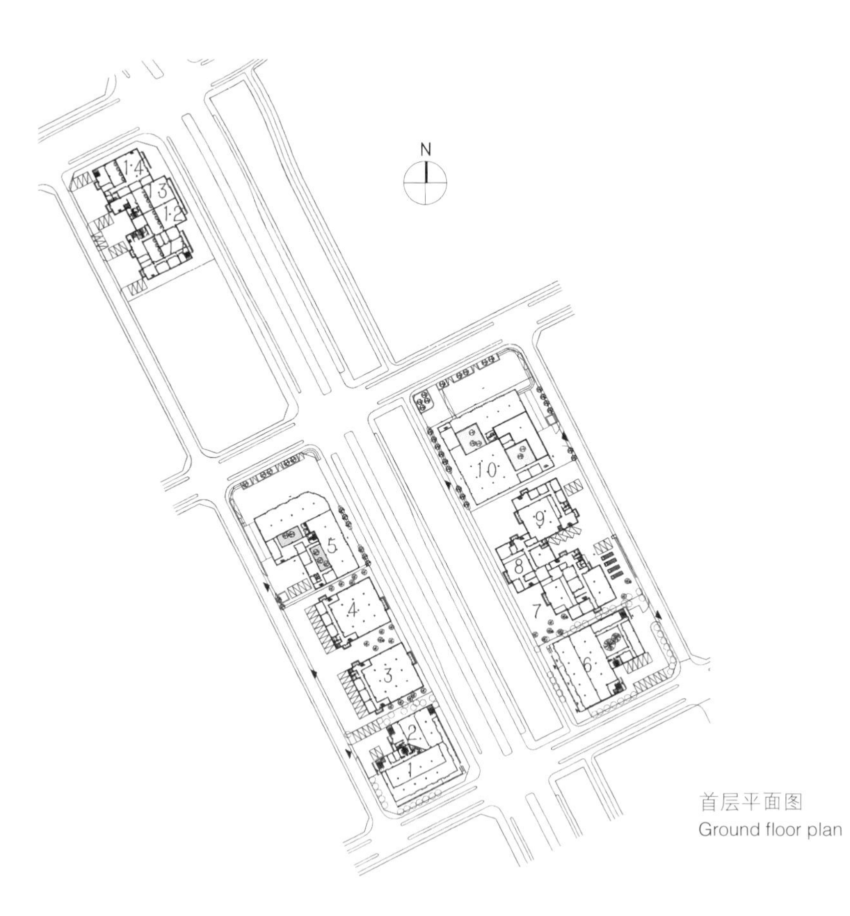

首层平面图
Ground floor plan

1 中国工商银行 ICBC
2 中国人民财产保险 PICC
3 中国银行 Bank of China
4 中国人寿保险 China Life Insurance
5 中国电信 China Telecom
6 中国农业银行 Agricultural Bank of China
7 中国邮政 China Post
8 中国人民银行 People's Bank of China
9 邮政储蓄 Postal Savings Bank of China
10 农村信用合作联社 Rural Credit Union
11 绵阳市商业银行 Mianyang Commercial Bank
12 北川富民村镇银行 Fumin Bank of Beichuan
13 中国农业发展银行 ADBC
14 兴业银行 Industrial Bank

BANK OF CHINA

北川经济开发区服务中心 Service Center of Beichuan Economic Development Area

援建：山东省烟台市

设计：烟台市建筑设计研究股份有限公司

施工：烟建集团有限公司

用地面积：23 867m²

建筑面积：15 315m²

服务中心 5 880m²

职工公寓 9 435m²

Aided by:
Yantai, Shandong Province

Designer:
Yantai Architectural Design and Research Co., Ltd.

Contractor:
Yanjian Group Co., Ltd.

Site area: 23,867m²

Building area: 15,315m²

Service Center: 5,880m²

Staff Apartments: 9,435m²

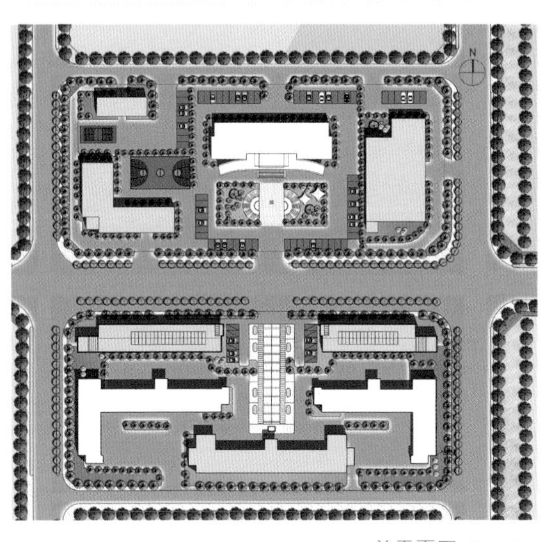

总平面图 Site plan

经济开发区服务中心位于开发区的西侧中部，由城市道路将其分隔为两部分。园区服务中心位于北侧地块，两座平面为L形的职工公寓则设于南侧地块。建筑单体设计主要体现工业园区简洁明快的特点，服务中心以现代风格为主，顶部通过对"白石"符号的抽象，体现了地域特色。公寓的形式则更具地域特点，同时采用了有利于自然通风和采光的单廊布局。

The Service Center is located in the middle of the west side of Beichuan-Shandong Industrial Park, which is divided into two parts by a town road, of which, the service center is on the north side of the site, while two staff apartment buildings in an L-shaped layout are set on the south side. The design for single buildings mainly reflects simple and lively features of the industrial park in a modern style, and the tops show the local features by applying the symbol of "white stones". The apartments are more featured with local elements and adopt a single corridor layout for natural ventilation and lighting.

路网规划概况
Road Network

北川新县城路网规划设计坚持以人为本，鼓励居民的绿色出行，全面建立人性化的交通体系。

小城镇尺度的路网规划

把握小城镇的城市尺度，避免照搬大城市的交通发展模式。规划构建"高密度、窄道路、小街坊"的整体路网模式。干路红线以20m为主，核心区道路间距原则上不超过200m，道路占地比例为17%，干路网密度7km/km²，路网整体密度高达10km/km²，从而在较少增加道路用地的情况下，显著提高道路网络的密度，提高街坊的可达性。

绿色交通主导的路网设计

规划倡导绿色交通主导的设计理念，以慢行交通优先为设计原则。将生活区划为稳静交通区，限制机动车速度，机动车速度不超过40km/h。在道路资源分配上，慢行交通的面积占全部道路面积（不包括道路绿化带）的一半，慢行通道的密度达到了17km/km²，慢行通道的密度高于机动车通道的密度。考虑到居民需求的多样性，慢行通道还包括健身、游览等多种功能。

宜人的细部设计

道路横断面方案突破仅以道路等级为出发点、以规范和标准为依据的传统做法。规划充分考虑道路功能、交通组织和交通需求的差异，分别做出多种类型的道路横断面。交通干路保证双向3～4车道、居住区干路提供路侧临时停车的空间、服务功能为主的干路为慢行交通留出更多的空间。所有干路快慢交通之间严格分离，避免机动车对慢行交通的干扰。将自行车与步行道设置在同一个平面上，采用不同的铺装进行区别，保证慢行交通空间的灵活性。在较大交叉口设置行人过街安全岛，保证慢行交通安全。失单适当减少交叉口转弯半径，限制机动车转弯速度，确保行人交通安全。尽量将公交车站设置在交叉口附近，保证公交乘客的最大便利等。（戴继锋 执笔）

The road network planning and design of the new county seat of Beichuan adheres to the human-oriented principle and encourages the residents to take eco-firendly means of transportation to establish a human-based traffic system.

Road network planning with a dimension of a small city

The road network planning should adopt the dimension of a small city, and transportation development mode of large cities should be avoided. The planning establishes an integrated road network mode of "high density, narrow road and small neighborhood". The site boundary of the trunk road is mostly 20m, and the distance between roads in core area should be no more than 200m in principle. The road-land ratio is 17%, trunk road network density is 7km/km² and the overall density of the road network is as high as 10km/km², so that the road network density can be significantly increased without increasing land for roads by large margin, and neighbourhood will be more accessible.

Green transport-oriented road network design

The planning advocates a concept of green transport-oriented road network design and making priority to non-motorized transportation as the design principle. The residential areas are designed as stable and quite transportation area in which the speed of motor vehicles is restricted at no larger than 40km/h. In terms of road resources allocation, the non-motorized transportation area should account for half of the total road area (excluding the area of road green belt), the density of non-motor vehicle passages should be 17 km/km², higher than that of the motor vehicle passage. The non-motor vehicle passages also function as a place for fitness and sight-seeing to satisfy multiple demands of the residents.

Pleasant detail design

Traditionally, road cross section scheme begins with road grade classification and takes the codes and standards as design criteria; however, in this road network planning, road cross sections of various types are adopted by considering differences among road functions, traffic organization and traffic demand. Trunk roads should be two-way and have 3-4 lanes; trunk roads in the residential area should be equipped with temporary parking space; trunk roads with main function of serving should leave more space for non-motorized transportation. Motorized transportation and non-motorized transportation on all trunk roads should be strictly separated from each other to avoid interference of motor vehicles on the non-motorized transportation. The bicycle and pedestrian passages are set on the same plane, and adopted with different pavement methods to ensure the flexibility of the non-motorized transportation. Pedestrian islands should be set at major road crossings to ensure the non-motorized transportation safety. The turning radius of road crossings should be properly reduced and the turning speed of motor vehicles should be controled to ensure pedestrian safety. Bus stations should be set at locations as close as possible to the road crossings to facilitate bus passengers. (by Dai Jifeng)

公交站亭、红绿灯、路灯、路牌等设施，也体现了设计中的风格原则 The design of bus station, traffic lights, street lights and street nameplates follows the principle of architectural style.

交通干路与居住区干路交接处 The intersection of arterial road and sub-arterial road in residential quarter

交警用车（孙彤 摄）Police Patrol Electro Mobile (Photo by Sun Tong)

永昌大道 Yongchang Avenue

永昌大桥（中规院供稿）Yongchang Bridge (by CAUPD)

辽宁大道（中规院供稿）Liaoning Avenue (by CAUPD)

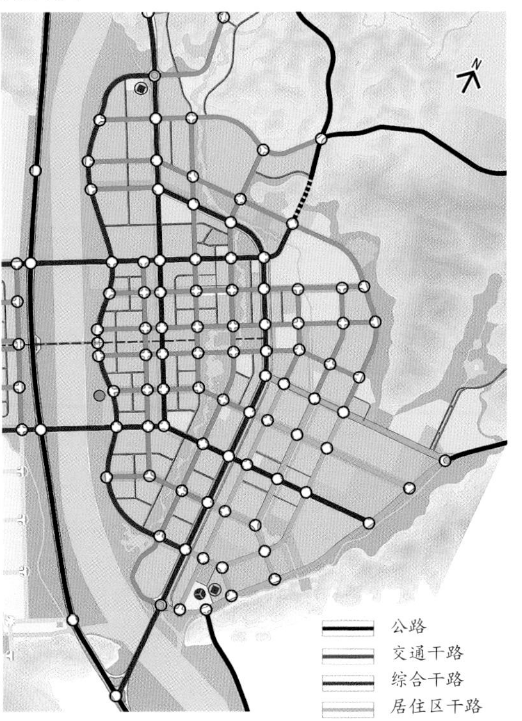

公路
交通干路
综合干路
居住区干路
工业区干路
滨河路
隧道
步行专用路
支路
平路平面交叉口
综合交通枢纽
旅游停车场
公交综合场站

北川新县城道路功能布局图 The road network functional division of the new county seat of Beichuan

永昌大桥 Yongchang Bridge

援　建：山东省济南市

方　案：上海城西建筑工程勘测设计院

施工图：济南市政工程设计研究院有限责任公司

施　工：济南城建工程公司

济南黄河路桥工程公司

主桥长度：268m

宽　　度：34m　双向六车道

Aided by:
Jinan, Shandong Province
Scheme design:
Shanghai Chengxi Architecture Engineering Consultants Institute
Construction drawings:
Jinan Municipal Engineering Design & Research Institute Co., Ltd.
Contractors:
Jinan City Construction Engineering Company
Jinan Yellow River Bridge Engineering Company

Length of main bridge: 268m
Width: 34m, two directions with 6 lanes

永昌大桥位于新县城最南端，跨越安昌河，连接新县城主要干道永昌大道，是北川通往绵阳、安县的交通要道，地理位置较为重要。主桥结构为五跨拱梁结合的连拱桥，设计为斜桥，尽可能减小阻水面积，利于防洪，防洪标准为百年一遇。桥身采用混凝土变截面连续梁，底面形成连续优美的弧线，简洁完整，经济实用，符合其主要满足交通需求的特点。是新县城工艺最为复杂、工作量最大、施工难度最大的桥梁。

Spanning the Anchang River, the Yongchang Bridge lies on the southernmost end of the new county seat and connects the Yongchang Avenue. It enjoys an important geographical position and is a vital traffic path from Beichuan to Mianyang City and An County. The bridge is designed as a skew bridge to minimize the area of water barrier for easy flood control with a standard of a 100-year recurrence. Moreover, the bridge adopts concrete continuous beam with variable cross-section, and its bottom surface forms a continuous and graceful arc, which is concise, complete, economical and practical, and conforms to the traffic demands.

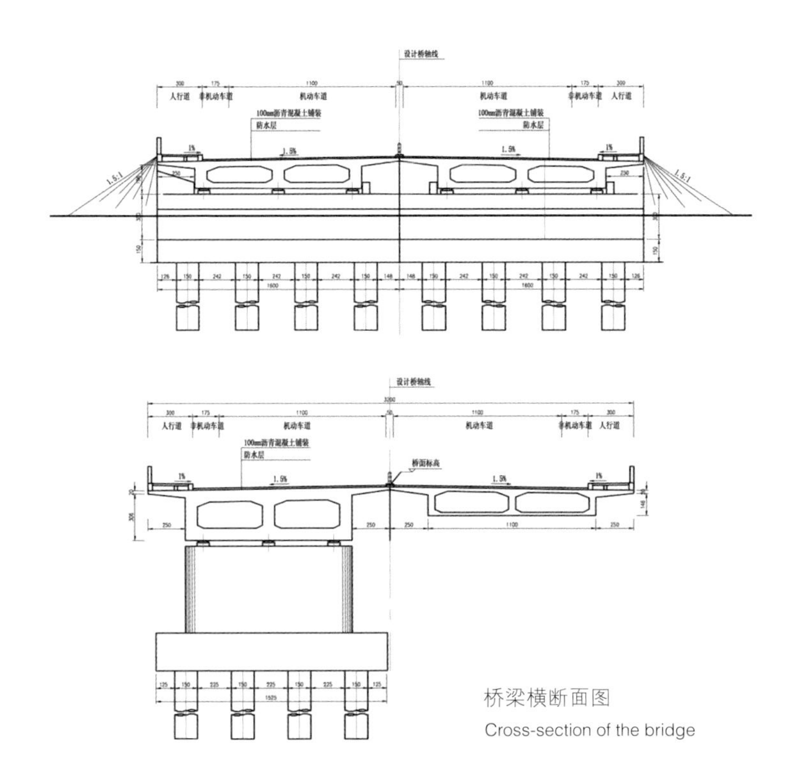

桥梁横断面图

Cross-section of the bridge

西羌北桥 Xi Qiang Bei Qiao Bridge

援　建：山东省潍坊市

设　计：潍坊市市政工程设计研究院有限公司

施　工：潍坊市市政工程有限公司

主桥长度：205m

宽　　度：30m　双向四车道

Aided by:
Weifang, Shandong Province
Designer:
Weifang Municipal Engineering Design & Research Institute Co., Ltd.
Contractor:
Weifang Municipal Engineering Co., Ltd.

Length of main bridge: 205m
Width: 30m, two directions with 4 lanes

西羌南桥 Xi Qiang Nan Qiao Bridge

援　建：山东省淄博市
方　案：淄博市规划设计研究院
施工图：淄博市建筑设计研究院
施　工：山东鲁中公路建设有限公司

主桥长度：197m
宽　　度：30m，双向四车道

Aided by:
Zibo, Shandong Province
Scheme design:
Zibo Urban Planning Design Institute
Construction drawings:
Zibo Architectural Design & Research Institute
Contractor:
Shandong Luzhong Road Construction Co., Ltd.

Length of main bridge: 197m
Width: 30m, two directions with 4 lanes

汽车客运站 Bus Station

方　案：四川省建筑设计院

施工图：信息产业电子第十一设计研究院

施　工：四川安汉建设有限公司

用地面积：21 294m²

建筑面积：3 334m²

停 车 位：113个

Scheme design:
Sichuan Provincial Architecture Design Institute
Construction drawings:
The Eleventh Design & Research Institute of IT Co., Ltd.
Contractor:
Sichuan Anhan Construction Co., Ltd.

Site area: 21,294m²
Building area: 3,334m²
Parking space: 113

汽车客运站位于新县城南端，城市干道永昌大道东侧，是北川通往绵阳、安县的南大门。地块呈三角形，根据功能分设客车停车发车、乘客集散和城市公交三个区块，避免了流线的混杂。邻近道路交叉口处则布置绿地，减少对周边景观和城市道路的干扰。建筑单体平缓舒展，以环抱式的折线布局，呈现出欢迎的姿态。建筑体量东端高起，既模仿羌族四角碉楼的形式，也是对整体形象的强调。

The Bus Station, being a gate of the new county seat leaing to Mianyang City and An County, is located on the south end of the town, and on the east side of the Yongchang Avenue. This triangular site is divided into three areas by functions, namely bus arrival and departure, passenger area and town bus area to separate flows. The green space near the road intersection is set to reduce interference to surrounding landscape and roads. On the whole, the building is in an embracing layout as if it welcomes all passengers. The building volume appears high in the east end, which imitates the form of square Qiang turret, and is also a highlight of the overall image of the station.

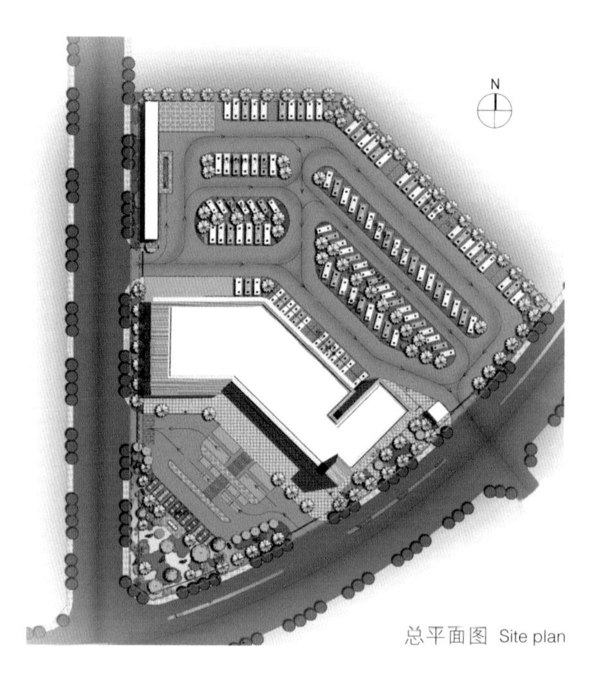

总平面图 Site plan

园林绿地系统规划概况

Green Spaces

城市绿地景观系统是城市功能空间的有机组成部分。园林绿地系统规划设计全面贯彻北川新县城建设的"安全、宜居、特色、繁荣、文明、和谐"标准和把北川新县城建成"城建工程标志、抗震精神标志和文化遗产标志"的目标。

突出绿地景观作为城市"绿色基础设施"的作用，并在灾后重建中置于"先行建设"的地位。

1. 合理安排城市防灾和公园绿地，在满足新县城300m公园服务半径的同时，也使城市防灾避险空间分布更均匀更充分。

2. 将城市防洪排涝功能与水系景观相结合，实现功能、美学和游憩的统一。新川河虽以城市排水为主，但毗邻主要商业街，景观设计突出雨水收集与城市滨水空间使用；永昌河水系丰富，景观设计考虑水质生态净化和雨水综合利用等要求。

3. 保证城区绿量，采用较大规格乡土苗木，迅速形成林荫蔽城的效果，建成区的绿地面积占建设用地的17%，人均18m^2，保证新县城在使用之始就能有一个良好的生态环境。

充分利用场地的自然环境条件，构筑新县城的景观生态格局。

北川新县城位于安昌河河谷盆地之中，自然景观呈现"群山环抱，一水中流"的特点，规划以周边山体和原有农田灌溉沟渠为基础，形成"一环、两带、多廊道"的网络状景观生态格局。

1. 一环——环城山体与郊野公园绿环。在保护现有山林植被的基础上，进行生态恢复和景观提升，并通过山地郊野公园建设，把生态防护与风景旅游功能结合起来。

2. 两带——两条生态景观游憩绿带。针对城市静风频率较高的特点，把城市绿地相对集中地布置为带状并顺应城市主导风向，形成永昌河和安昌河两条大型的生态通风廊道。生态廊道内兼有居民健身游憩设施，方便居民使用。

3. 多廊道——多条连水通山绿廊。绿廊建设与城市干道和水系空间相结合，形成4条南北纵向连水绿廊、8条东西横向通山绿廊，各条廊道兼有休闲游憩、景观和生态通风等多种功能。

强调"生活融于绿色"的理念，营造丰富多彩的户外绿化活动空间，增加居民交往，抚平灾后创伤。

1. 考虑到城市用地紧张的特点，将居住小区（组团）绿地与城市公园结合，以提高园林绿地的景观质量和管护水平，同时也有利于减少居民的物业负担。

2. 保护原场地古桥、鱼塘、树林、建（构）筑物等遗存，并将之组合到新的景观设计中，突出了景观的乡土性，场地的记忆性以及景观的归属感。

通过景观空间和小品，体现羌族风貌和"抗震"主题，丰富新县城文化展示途径。

1. 以抗震救灾和灾后重建的过程为主线，通过静思园、英雄园、幸福园三个特定空间，展示了伟大的抗震救灾精神。

2. 在友谊园中，以齐鲁文化元素和羌族元素的建筑小品为载体，通过自然湖面的联系，体现了"川鲁情谊深似海"的景观意境。

3. 景观小品设计注重羌族元素和地方材料的运用，形成景观特色。（束晨阳 执笔）

Urban green space landscape system is an integral part of urban functional space, so its planning and design should follow the principle of building a "safe, comfortable, vigorous, featured, civilized and harmonious" new county seat of Beichuan to realize the goal of building the county into "a symbol of urban construction, a symbol of quake relief spirit and a symbol of cultural heritage".

Highlighting the role of green space landscape as "green infrastructure" and "pioneer construction" in post-quake reconstruction

1. Rationally arrange urban disaster prevention and avoidance space and park area, so that it meets the demands of 300m park serving radius and makes urban disaster prevention and avoidance space be more evenly distributed;

2. Combine urban flood control and drainage system with the waterscape to integrate functions, aesthetic effects and recreation. The Xinchuan River is mainly designed for urban drainage, but as it is adjacent to the major commercial street, rainwater collection and utilization of urban waterfront space should be stressed in its landscape design; given the rich water system of the Yongchang River, water quality eco-purification and comprehensive rainwater utilization and other requirements should be taken into account in its landscape design;

3. Adopt native seedlings of larger size to ensure green rate in urban area and realize the effect of trees shading the whole county. The green area of the completed area should be 17% of that of the land for construction, or a per capita green area of 18m^2, so that the new county has a good eco-environment once it is put into use.

Building landscape ecology pattern of the new county seat by fully utilizing the natural conditions of the site.

Since the new county seat locates in the Anchang River valley basin and is surrounded by mountains with river running through

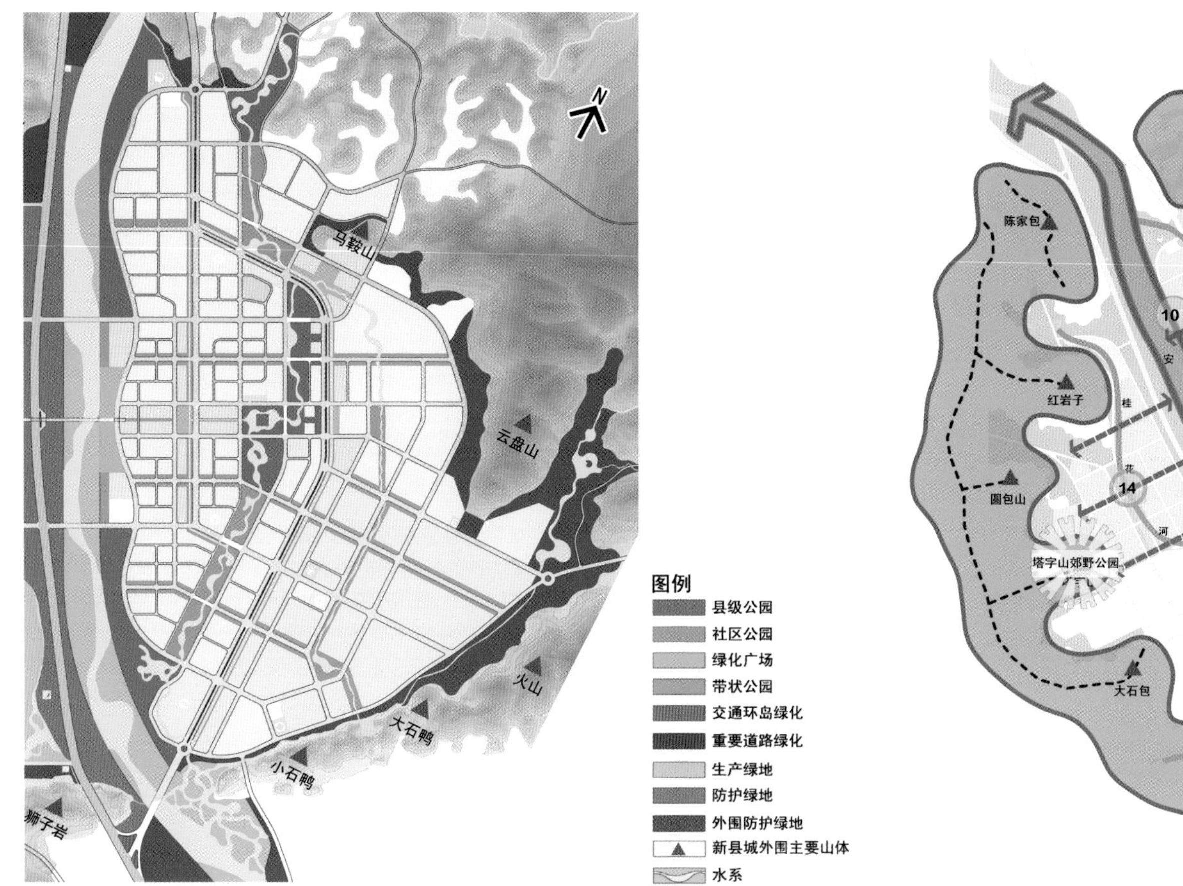

北川新县城绿地结构规划图

Planning of the green space structure in the new county seat of Beichuan

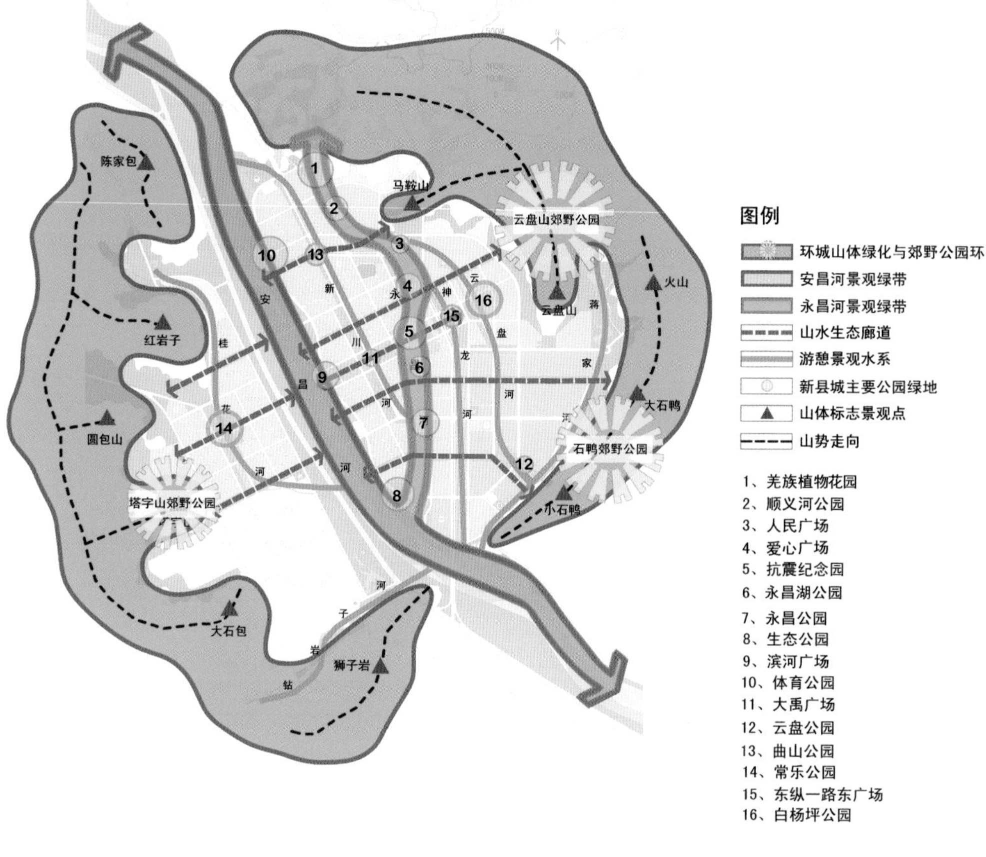

北川新县城绿地系统规划图

Planning of the green space system in the new county seat of Beichuan

it, by utilizing the surrounding mountains and the original channels and ditches for agricultural irrigation, the planning is made to form a network-shaped landscape ecology pattern featuring "one circle, two belts, and multiple corridors".

1. One circle—green circle made up of mountains around the county and suburb park. Restore ecology and improve landscape while protecting the existing forests and vegetation. Combine the functions of ecological protection and scenic tourism by setting up country parks in hilly areas;

2. Two belts—two ecological landscape recreation belts. Considering the high frequency of static wind in county, urban green space should be arranged centrally in striped shape along with the dominant wind direction, forming two large biological ventilation corridors of the Yongchang River and Anchang River. The eco-corridors are equipped with fitness and leisure facilities for the convenience of the residents.

3. Multiple corridors—several green corridors connecting mountains and water. The green corridor should be constructed by considering the urban trunk roads and water space. There are four south-north green corridors connecting water system and eight east-west green corridors connecting mountains, and each corridor serves as a space for leisure, landscape, eco-protection and ventilation.

Stressing the concept of "living in green environment", creating colorful outdoor green activity space, enhancing the communication among residents and easing the trauma of the 5.12 Wenchuan Earthquake

1. Given that the urban land is in short supply, the way of combining residential quarter (cluster) with urban park can improve the landscape quality and maintenance level of gardens and green space, and it also helps to reduce the property management burden of the residents;

2. Protect remains such as ancient bridges, fish ponds, forests, buildings (structures) in the original site, and incorporate them into the new landscape design to highlight the localism and senses of belonging of the landscape as well as memory of the site.

Adopting landscape space and pieces to reflect the Qiang style and feature and the "quake-relief" theme, and diversify way of exhibiting culture of the new county seat.

1. The process of disaster relief and post-disaster reconstruction is taken as the main thread, and great disaster relief spirit is showcased through three specific space, they are, the Meditation Garden, the Hero Garden and the Happiness Pavilion;

2. By taking the architectural pieces with the Qilu cultural elements and the Qiang elements as the carriers, and through contact with the natural lake, the design of the Friendship Garden reflects the profound friendship between the Sichuan and Shandong people;

3. The Qiang elements and local materials are used in landscape pieces design so as to form distinctive landscape. (by Shu Chenyang)

安昌河景观工程 Landscaped Green Belt along Anchang River

援　建：山东省潍坊市、滨州市

设　计：山东省水利勘测设计院

中国城市规划设计研究院

北京北林地景园林规划设计院

施　工：潍坊昌大建设集团有限公司

青岛绿地生态技术有限公司

天津市绿化工程公司

江都市园林工程有限公司

山东滨州城建集团公司

济南华泰园林绿化公司

面　积：335 000m^2

Aided by:

Weifang and Binzhou, Shandong Province

Designers:

Water Conservancy Survey and Design Institute of Shandong Province

China Academy of Urban Planning & Design

Beijing Beilin Landscape Architecture Institute

Contractors:

Weifang Changda Construction Group Co., Ltd

Qingdao Green Area & Ecological Technique Co., Ltd

Tianjin Greening Engineering Company

Jiangdu Landscape Engineering Co., Ltd

Shandong Binzhou Urban Construction Group Ltd

Jinan Huatai Landscape Greening Co., Ltd

Area: 335,000m^2

安昌河东岸景观带毗邻城市。设计以带状绿化空间强调对城市活动的服务，突出运动健身和生态科普功能，包括南部生态公园、中部滨河绿化广场和北部体育公园三个组成部分。强调大尺度整体植物景观效果。安昌河西岸景观带功能为城市过境道路的绿化隔离带，以植物景观为主，通过自然风景林的营造，兼顾防护与美化效果。

The east bank of the Landscaped Green Belt along Anchang River is close to the new county seat. The striped green space design stresses the concept of serving town's activities and highlights the function of fitness, biological knowledge popularization, and it is composed of the south ecological park, riverside green plaza in the central area and north sports park. Overall effect of large size plant landscape is stressed. The west bank of the landscaped green belt serves as the greenbelt of urban throughway with plants on both sides. It creates natural landscape forest while considering protection and beautification effect.

安昌河东岸的羌笛广场 The Qiang Flute Plaza along the east bank of the Anchang River.

南部生态公园中的儿童游戏场

Children's playground is arranged in the south ecological park.

景观带中为居民设置了数十处活动场地

Dozens of activity venues in the landscaped green belt are built for residents.

将原场地鱼塘等遗存构筑物与新的景观设计结合

Remaining structures in the former plot such as fish pond are incorporated into the new landscape design.

永昌河景观工程 Landscaped Green Belt along Yongchang River

援　建：山东省青岛市
设　计：中国城市规划设计研究院
　　　　中国风景园林规划设计研究中心
　　　　北京北林地景园林规划设计院
施　工：青岛第一市政工程有限公司
　　　　青岛城建集团有限公司
　　　　青岛花林实业有限公司
　　　　青岛市黄岛园林绿化工程有限公司

面　积：457 600m^2
长　度：3 490m
宽　度：80～200m

Aided by:
Qingdao, Shandong Province
Designers:
China Academy of Planning and Design
China Research Center for Landscape Planning and Design
Beijing Beilin Landscape Architecture Institute
Contractors:
Qingdao NO.1 Municipal Engineering Co., Ltd.
Qingdao Urban Construction Group Ltd
Qingdao Hualin Shiye Co., Ltd.
Qingdao Huangdao Landscaping & Afforesting Co., Ltd.

Area: 457,600m^2
Length: 3,490m
Width: 80~200m

永昌河景观带贯穿新县城南北，设计以展示新县城形象、体现抗震精神、继承和弘扬羌族文化、丰富市民文化生活为功能定位。依据景观带与周边城市用地布局关系，通过一系列开放性公园绿地和景观小品的营造，成为地域特色鲜明的综合性城市生态景观（公园）带。

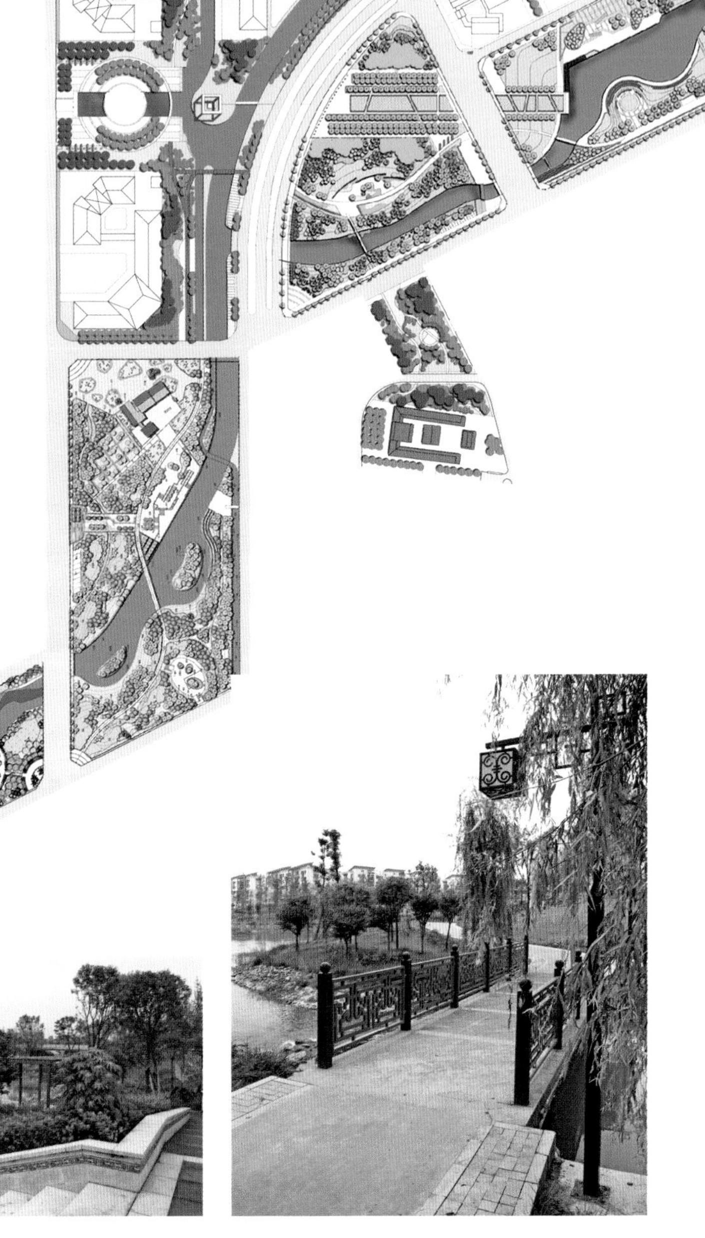

永昌河景观带总平面图
Site plan of Landscaped Green Belt along Yongchang River

The Landscaped Green Belt along Yongchang River passes through the new county seat from south to north. The functions of the belt are to demonstrate the image of the new county, reflect disaster relief spirit, inheriting and carrying on the Qiang culture, and enrich the cultural life of the urban dwellers. In accordance with the relationship between the landscape belt and the surrounding urban land layout, the Yongchang River Landscape Belt becomes a comprehensive urban biological landscape (park) belt with distinctive regional features through a host of open parks, green space and landscape pieces.

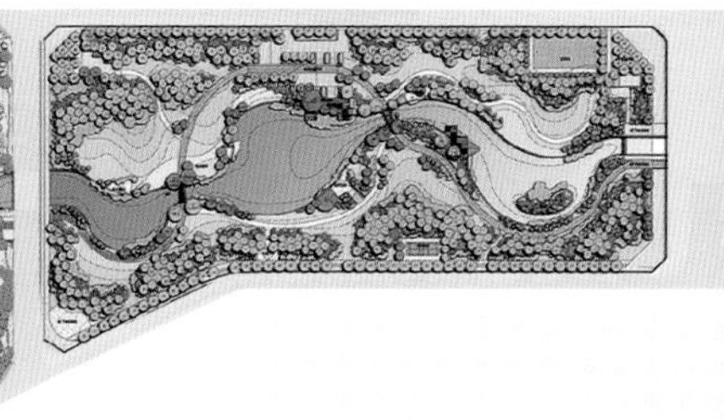

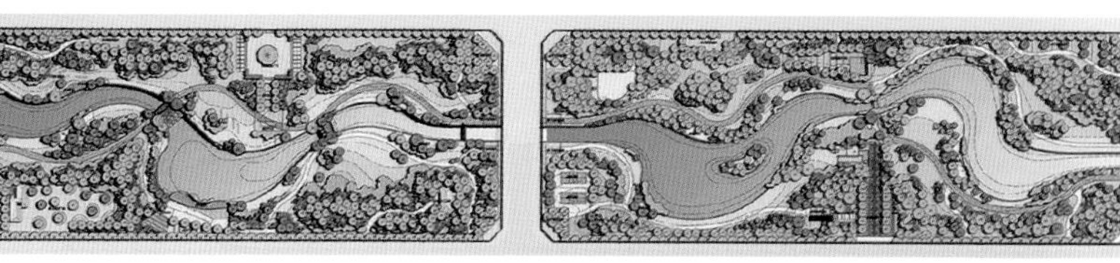

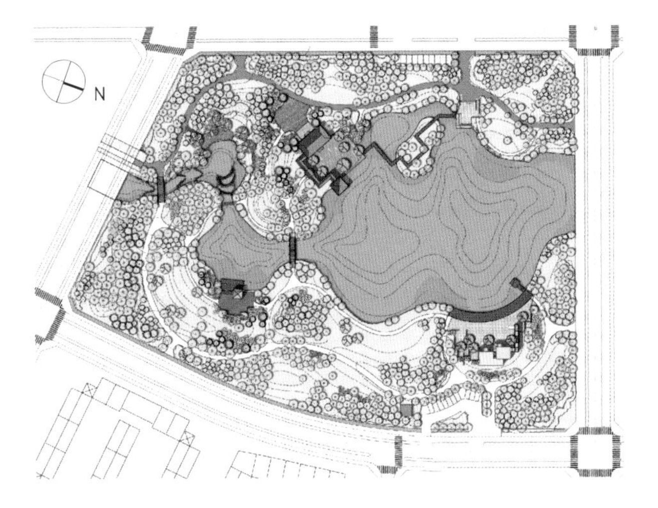

友谊园总平面图 Site plan of Friendship Garden

友谊园是永昌河景观带最大、最集中的一块用地，面积2.9ha，以齐鲁文化元素和羌族元素的建筑小品为载体

The Friendship Garden, is the largest and most centralized plot of the Yongchang River Landscape Belt. It has a total area of 2.9ha, focusing on the architectural elements of Shandong culture and Qiang ethnic culture.

鲁韵庭院赋

新川河景观工程 Landscaped Green Belt along Xinchuan River

援　建：山东省淄博市

设　计：中国风景园林规划设计研究中心

济南市市政工程设计研究院

施　工：山东天齐置业集团股份有限公司

面　积：59 700m²

长　度：2 900m

宽　度：23.5m

Aided by:
Zibo, Shandong Province

Designers:
China Research Center for Landscape Planning and Design
Jinan Municipal Engineering Design & Research Institute Co., Ltd.

Contractor:
Shandong Tianqi Industry Group Ltd

Area: 59,700m²
Length: 2,900m
Width: 23.5m

局部驳岸采用亲水平台、亲水台阶，增加市民观赏、休闲的趣味

The Qiang elements and local materials are used in landscape piece design.

新川河景观带设计利用场地原有灌渠的基础，结合毗邻城市商业、居住街区的特点，突出整齐大气，节奏多变的线性景观肌理，通过丰富的亲水平台、林荫场地、水生植物种植池等景观元素的营造，成为城市中充满活力和趣味的步行绿廊。

The Landscaped Green Belt along Xinchuan River neighbours the commercial and residential street blocks, so the original irrigation ditch in the site are utilized in the design to highlight the linear landscape texture which is regular and grand, with changing rhythm. It becomes a green pedestrian passage full of dynamics and fun in the county through the creation of rich water-intimate platform, tree shades and planting bed of aquatic plants.

景观小品设计注重羌族元素和地方材料的运用

Water platforms and steps are arranged locally on the revetment for residents to appreciate in leisure time.

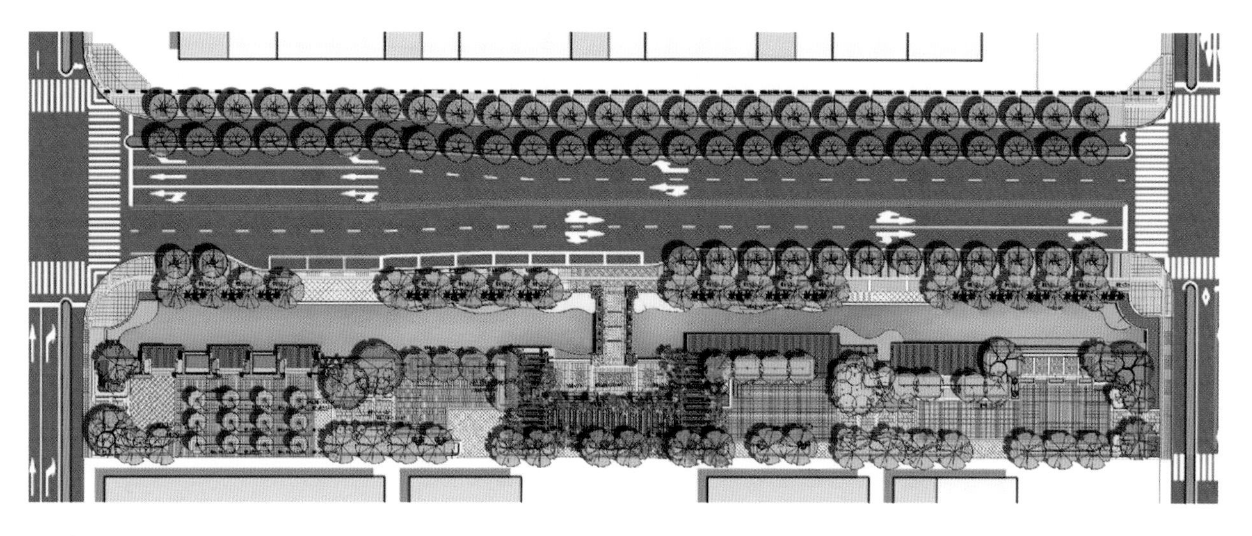

新川河商业区段平面图

Landscape design of business area along the Xinchuan River

北川山东产业园

Beichuan-Shandong Industrial Park

北川山东产业园区是震后建立的省级北川经济经济开发区“一区多园”中的核心园区，是“再造一个新北川”的重要经济引擎，是促进就业、生态和谐、有地方特色的示范性的产业基地。该园区位于新县城的东南角，距绵阳市区约25km。园区面积约120.7ha。

产业遴选确定基本的原则包括：

·扩大就业为优先目标，鼓励劳动密集型产业和为城镇服务的产业发展。

·以集约用地为重要要求，产业项目应满足一定的投资和产出强度，强调投资与用地投放直接挂钩。

·因地制宜，突出地方资源特色，有针对地扶持体现民族地方特色的产业。

·促进北川县工业化程度，承接区域产业转移为战略目标，争取引入龙头大型骨干企业。

·严禁高污染、高耗能产业的入驻。

规划确定六大产业及其空间布局，分别为：北川特色产业、电子信息及新材料、机械设备制造、新型建材、农特产品加工和服装纺织。

根据国家相关规范和总体规划，产业园区的开发总量控制在90万～100万平方米左右，其中工业建筑开发量约60万～70万平方米。工业园区以低层厂房为主，沿永昌大道一侧集中布置办公楼和配套服务设施，建筑以多层和小高层为主。

目前山东产业园已有24家企业正式入驻建设，待全部开工生产后园区内的工业企业可为新县城及全县提供一万多个就业岗位，大大促进灾后重建的经济发展和减轻新县城的就业压力，也给部分进城打工人员返乡就业提供了良好的机会。（蔡震、张晋庆、陈烨 执笔）

The Beichuan-Shandong Industrial Park represents a core park of the provincial-level Beichuan Economic Development Area which was established after the 5.12 Wenchuan Earthquake, an important economic engine for “Building a New Beichuan”, and a pilot industrial base with local characteristics for promoting eco-harmony and employment. Covering an area of 120.7ha, this park is located in the southeast corner of the new county seat of Beichuan, and about 25km away from downtown Mianyang.

Basic principles for selecting industries in the park:

·Take employment expansion as the priority, and encourage development of labor-intensive industries and industries serving the town;

·Based on the major requirement of intensified land use, the industrial projects should satisfy certain investment and output intensity, stressing direct linkage between investment and land application;

·Adopt measures to local conditions, highlight local resource features, and give special support to industries with local and ethnic characteristics;

·Promote industrialization of Beichuan county, take regional industrial transfer as the strategic target, and introduce leading backbone enterprises;

·Industries with high pollution and high energy consumption are strictly prohibited from entering the park.

The planning defines six industries and spatial layouts which are Beichuan characteristic industry, IT and new materials, machinery and equipment manufacturing, new building materials, special agricultural product processing and garment & textile.

In accordance with relevant national codes and the general planning, the total development volume of the Beichuan-Shandong Industrial Park is around 900,000 to 1,000,000m^2, of which, the area of the industrial buildings totals 600,000 to 700,000m^2. The industrial park is mainly composed of low-rise workshops, and office building and supporting service facilities, mostly multi-storey and medium-rise buildings, are built along the Yongchang Avenue.

Till now, altogether 24 enterprises have been in progress of construction in Beichuan-Shandong Industrial Park, and when all enterprises in the park commence construction, they can offer more than 10,000 job opportunities for the new county seat and the whole county, which will significantly promote post-disaster economic growth, relieve employment pressure for the new county seat, and attract some migrant workers in big cities to come back to work.

(by Cai Zhen, Zhang Jinqing and Chen Ye)

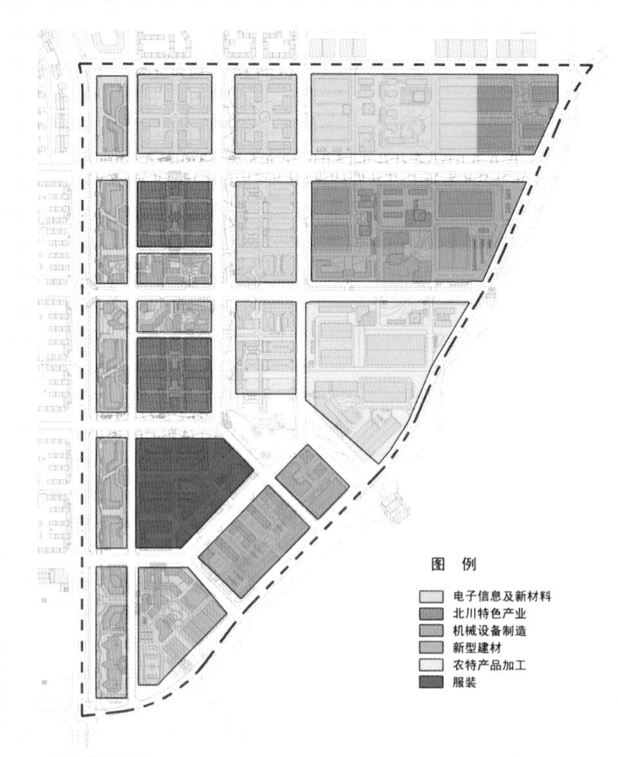

产业分区图 Industrial zoning

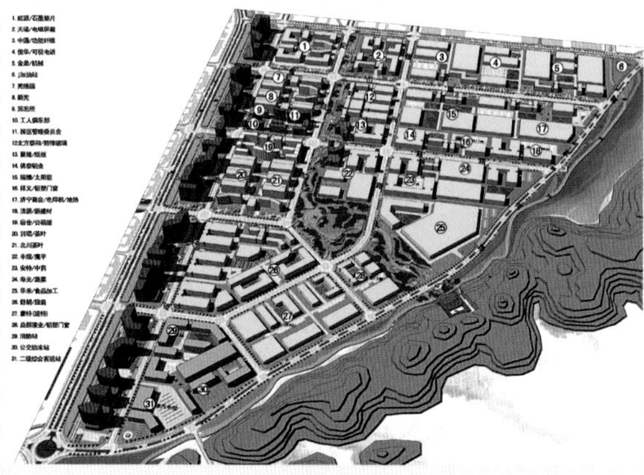

园区建筑布局意向 Architectural layout idea

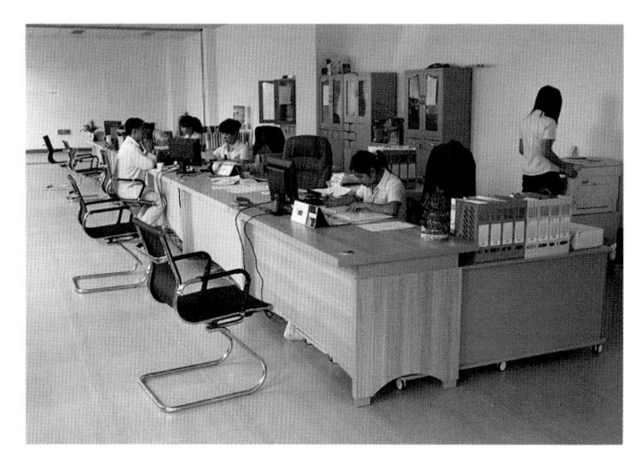

工业园区的部分厂房已经投入使用　Several facotries of the industrial park are in process of production（孙彤 摄 Photo by Sun Tong）

农业科技示范园

Agricultural Technology Demonstration Garden

农业科技示范园及农产品交易市场项目位于新县城西北部分，安北公路西侧，与安昌河东岸的城区有便捷的联系，可通过安北公路与北川县域广大农村地区和绵阳城区联系。

该项目用地南部是特色农产品市场，围绕交易大厅建设配送、储藏中心等配套设施，并在其周围分布了精品展示店面。北部示范区包括核心的展示内容（展示厅、生产温室、示范温室、研发温室、研发中心、培训中心），还包含了北部的高山植物低海拔驯化试验展示基地。西北部是园区的生活配套区，包含了员工公寓、专家公寓及相关配套设施。这里紧邻研发中心，环境相对幽静，结合溪渠利用地形营造水景，构建优美的生活中心。

为保护周边山体自然背景，建筑以低层为主，并形成层次丰富的城市天际线。

用地面积：402 100m^2

建筑面积：94 700m^2

Site area: 402,100m^2

Building area: 94,700m^2

The Agricultural Technology Demonstration Garden and the Farm Product Trading Market is located on the northwest of the new county seat, the west side of Anbei Avenue. It has a convenient traffic link with the east part of the town, while connecting the rural area of Beichuan County and Mianyang city easily through the Anbei Avenue.

The Farm Product Trading Market is seated on the south side, while the trading hall is enclosed by the service facilities such as the delivery center and the storage center. Several high-grand exhibition stores are also set around the hall. The north part of the project is the demonstration garden, which is composed of exhibition hall, production greenhouse, demonstration greenhouse, research greenhouse, research center and training center. An exhibition base of alpine plant low altitude domestication test is also located in the garden. The living area is on the northwest side of the site, adjacent to the research center. It contains stuff apartment building, expert apartment and relative service facilities.

To protect the surrounding natural mountains, most of the buildings are 1~3 storeys and composes diversified skylines.

北部示范区温室 The green houses in north demonstration garden（中规院供稿 by CAUPD）

总平面图 Site plan

农业科技示范园试验展示基地　Agricultural Technology Demonstration Garden Exhibition Base （孙彤 摄 Photo by Sun Tong）

后记

相信三年前我们很少有人知道北川这个小县城。那一场大地震却把世界的目光吸引到那个峡谷中的小城。从那一刻起，无数的人们在关心着北川，关注着新北川的建设。

作为震后唯一异地重建的县城，新北川从选址、规划、设计到实施都采用了不同寻常的建设模式。这中间不仅表现了援建者的爱心和责任，也摸索出了一套科学建设管理的经验。我们出版这本画册，一方面是歌颂，把一个漂亮的新北川呈现给关心她的人们；一方面是记录，把这短短的、不到三年的建设历程载入史册；一方面是思考，一个新城的建设一定有不少经验教训值得我们反思；还有一方面是祝福，无数援建者付出的劳动献给北川人民一个新的家园，新的生活！

看到这座美丽小城的一栋栋建筑和环境，无数新北川建设的参与者，会不时回忆起那些难忘的日日夜夜，回忆起那些难忘的淌着泪水或汗水的千千万万个面孔。显然，我们不能在这里展示所有那些令人感动的内容，我们只希望这座画册能引起大家的回忆和联想。

我们在这里应该感谢上至中央，下至北川县的各级领导对新北川建设的关心和支持，应该感谢中国建筑学会、规划学会各位专家的精心指导，应该感谢所有参与设计的规划师、建筑师和工程师们，更应该感谢山东省参与援建施工的工程技术人员和工人师傅们，他们辛勤的劳动将永远记录在这片土地上！

这本书呈现的是新北川建筑崭新的一瞬，新北川真实的生活刚刚起步，城市建设还将延续，让我们为她的新生而庆贺，更为她的成长而祝福！

崔　愷

2011.8

Afterword

It's true that three years ago before the 5.12 earthquake, few of us knew about such a small county as Beichuan. However, this earthquake drew all global attention to that small place in valleys, and from that moment on, numerous people have been caring about Beichuan and its construction in a new location.

As the only off-site rehabilitated county seat after the earthquake, the new county seat of Beichuan is applied with an unconventional aiding mode for its site selection, planning, design and construction, during which heart to heart care and responsibility of the aiding hands are demonstrated and a complete system of project management is also explored. On this occasion of publishing the book, we mean to celebrate and present a fresh new and beautiful town to those who are deeply concerned with the county; at the same time, we record the construction progress which is less than three years into the history. For one thing, it shows our introspection: the experiences and practice of building a new town deserve much deliberation; for another, it contains our blessing: Beichuan people are blessed with a new home and new life through continuous endeavor of the aiding hands.

Looking at the architecture, landscape and environment of this small and beautiful town, thousands of participants in building the new Beichuan will recall those unforgettable days and nights we worked there, and thousands of faces with tears and sweat we saw there. It's a pity that we cannot show all the moving memories, but with this book, we just expect to rekindle our reminiscence.

We are very grateful for the care and support of leaders of different levels to the rehabilitation of the new county seat, for meticulous instruction of experts from Architectural Society of China and Urban Planning Society of China, for all planners, architects and engineers involved in the design of new Beichuan, and for, in particular, those engineering technicians and workers for the aiding projects by Shandong Province whose industrious work will be forever recorded on this land.

This book is only a glance at buildings in the new county seat of Beichuan, while a real life there just begins and town construction will still continue. Let's celebrate her rehabilitation and bless her for well growth!

Cui Kai

August, 2011

图书在版编目（CIP）数据

建筑新北川/中国城市规划设计研究院，中国建筑设计研究院编著. —北京：中国建筑工业出版社，2011.10
ISBN 978-7-112-13648-3

Ⅰ. ①建… Ⅱ. ①中… ②中… Ⅲ. ①城市规划－建筑设计－北川羌族自治县－图集 Ⅳ. ①TU984.271.4-64

中国版本图书馆CIP数据核字(2011)第204353号

策　　划：张广源　张惠珍
责任编辑：孙　炼
责任校对：关　健　王雪竹

建筑新北川

Building A New Beichuan

中国城市规划设计研究院
中国建筑设计研究院　编著

中国建筑工业出版社出版、发行（北京西郊百万庄）
各地新华书店、建筑书店经销
北京方嘉彩色印刷有限责任公司印刷

开本：889×1194毫米　1/12　印张：20　字数：760千字
2011年10月第一版　2012年3月第二次印刷
定价：260.00元
ISBN 978-7-112-13648-3
(21408)